The Science Teachers' Handbook

Ideas and Activities for Every Classroom

Andy Byers
Ann Childs
Chris Lainé

The VSO ECOE Programme

Heinemann Educational Publishers
A Division of Heinemann Publishers (Oxford) Ltd
Halley Court, Jordan Hill, Oxford OX2 8EJ

Heinemann: A Division of Reed Publishing (USA) Inc.
361 Hanover Street, Portsmouth, NH 03801-3912, USA

Heinemann Educational Books (Nigeria) Ltd
PMB 5205, Ibadan
Heinemann Educational Boleswa
PO Box 10103, Village Post Office, Gaborone, Botswana

FLORENCE PRAGUE PARIS MADRID
ATHENS MELBOURNE JOHANNESBURG
AUCKLAND SINGAPORE TOKYO
CHICAGO SAO PAULO

Voluntary Services Overseas
317 Putney Bridge Road, London

First published by Heinemann Educational Publishers in 1994

British Library Cataloguing in Publication Data
A catalogue record for this book is available from the British Library

ISBN 0435 92302 1

Acknowledgements

The VSO ECOE Programme and authors thank all the VSO science teachers and their national colleagues for contributing the original material over the years from which this book was compiled, and also for trialling and commenting on early drafts of the text. Without their hard work, inventiveness and willingness to share ideas with others, this book would not have been written.

Particular thanks go to Hans Schmidt, Teacher Trainer, Erzhausen, Germany; Stephen Byers, Myers Grove School, Sheffield and Professor Ricardo Sanchez, Bolivia for contributing ideas to many parts of the text, and to Terry Allsop, Department of Educational Studies, Oxford University; Peter Fell, Science Adviser, British Council and Trevor Roach, Head of The Wilderness Centre for Environmental Education in Gloucestershire, for their advisory work.

We would like to thank Paul Ndurguru for providing a design for the safety goggles on page 9.

Designed and typeset by Susan Clarke, Reading, Berkshire
Illustrated by Willow
Printed and bound in Great Britain
The Bath Press, Avon

94 95 96 10 9 8 7 6 5 4 3 2 1

Contents

Introduction

Why has this book been compiled?

Teaching is a challenging and time consuming activity. Teachers are constantly looking for new ideas and practical work for science subjects. Many have to teach new or unfamiliar topics with limited time to plan and try out suitable activities. Pupils usually hope that science will offer exciting activities or experiments and science teachers face the challenge of meeting these expectations.

This book has been compiled by VSO to bring together successful practical ideas used by teachers all over the world. These ideas have been developed and adapted over many years by VSO teachers and their national colleagues working together in schools throughout Africa, Asia, the Caribbean and the Pacific. Based on this depth and breadth of experience, this text shows how to demonstrate science in action in clear and exciting ways, even when time and resources are limited.

In order to make the ideas in this book as practical and relevant as possible to the needs of teachers and teacher trainers, the book has been widely tested. Teachers in secondary and junior secondary schools, in workshops and in curriculum development units in more than 20 countries worldwide have trialled and commented on drafts and improved the final contents.

The ideas and activities in this book are presented to show what is possible and to encourage teachers to use them as starting points. They should be modified according to what is available and appropriate in local circumstances. This book is designed to be used as a resource alongside other materials, it is not a textbook.

Criteria for including ideas

These are the guidelines that have been used to select the material for inclusion, but not all criteria apply to all examples!

Each activity should

- clearly show the principle intended
- be used for more than one activity
- use commonly available materials, but not rely on imports
- be inexpensive, using few consumables or be re-usable
- be dismantleable
- be recyclable
- be storable
- build on initiatives already in practice in some countries, e.g. supply of science kits.

What does this book hope to achieve?

The book aims to be a useful resource for new and experienced teachers in countries throughout the world.

- To link the classroom with the community and use the science being practised in the community as a rich resource for delivering the science curriculum.
- To share ideas that have been successfully used all over the world for teaching science in a practical and active way.
- To show that 'classic' textbook experiments can be done without imported or expensive equipment.
- To inspire teachers to extend the variety of resources they draw on to teach science.

- To encourage teachers to think positively and creatively in making maximum use of local resources.
- To ensure that science teaching and learning is firmly based in everyday experience.

How to use this book

The Science Teacher's Handbook is best used as part of an approach to teaching using a scheme of work as the central planning document. It is only one resource out of several that can be used to make teaching varied, interesting and able to meet the needs of all students.

One example of a scheme of work is shown here, but many countries have a set format which should be used.

Scheme of work

	Syllabus	Learning objectives	Pupil activities
lesson 1	Term 3, week 4 The production of sound by a vibrating source. • properties of sound • simple treatment of pitch, loudness and quality.	By the end of the lesson the student should.... • know how different musical instruments make a sound. • know that sound can have a high or low pitch and loud or soft volume.	Observe a demonstration by musicians to show how a range of instruments produce sounds. • Show high notes on thin strings, under high tension and short tension. • Practical: Pupils work with musicians to test these ideas for themselves. • Homework: Students to collect materials to make instruments.
lesson 2		• understand that ideas (hypotheses) in science should be tested.	• Investigation: Pupils predict results of Experiment 1 (less able) or Experiment 2 (more able) and test their predictions.
lesson 3	Transmission of sound including speed.	• understand that sound is a wave of disturbed particles. • understand that sound travels in solids and liquids. • understand that sound cannot travel in a vacuum.	• Teacher demonstration and explanation of movement of sound through air. • Pupil activity: sound through a desk. • Teacher explanation of experiment using poster as visual aid.
lesson 4		• understand that sound has speed. • know that sound can be reflected.	• Listen to introduction to experiment and instructions. • Students carry out echo experiment using 2 bricks to make a 'clap'. • Homework: Write up experiment or use school library to find out about echo sounding.

Syllabus

The syllabus identifies what you should cover in your teaching scheme. You can add additional areas of interest relevant to the needs of your students.

Lesson plans

Each lesson should have a detailed plan – there is not enough room to fit all the information required into the scheme of work.

Learning objectives

These are sometimes given as part of the syllabus or teacher's guide – they help to clarify the purpose of the lesson.

...ources and ...erences	Assessment criteria	Teaching notes
...sicians ...udents' own ...struments ...e Science Teacher's ...ndbook pages 96 ...d 97. ...ge 97 Bamboo ...struments. ...ge 97 Hanging ...ects.	Oral evidence : (ask the students the following) • Where is sound coming from? • What is changing the sound? • What happens to sound when string is tighter/slacker? Suggested answers: • vibrating string • tightness of string • tighter gives higher sound	• Invite local musicians. Meet in advance to go through lesson plan etc. • Ask students to collect newspapers/magazines/articles on sound and music.
...ge 96 Changing ...ch. ...ge 96 Sound ...rough solids.		• Prepare The Science Teacher's Handbook pages 94 and 95 as extension work. • Ask local carpenter to make more metre rulers. • Make equipment for pages 96 and 97.
...ge 96 Drum ...rations ...ge 96 Sound ...rough solids.	Written evidence: Student writes statements about different effects, e.g. the shorter the vibrating length the higher the note.	• Prepare poster.
...ge 97 Speed of ...nd.		• Mention effect of wind and temperature. • Lesson outdoors – have other work available if weather unsuitable.

Assessment records

Information from exams, tests and class assessment is important for monitoring student progress and setting appropriate levels of work.

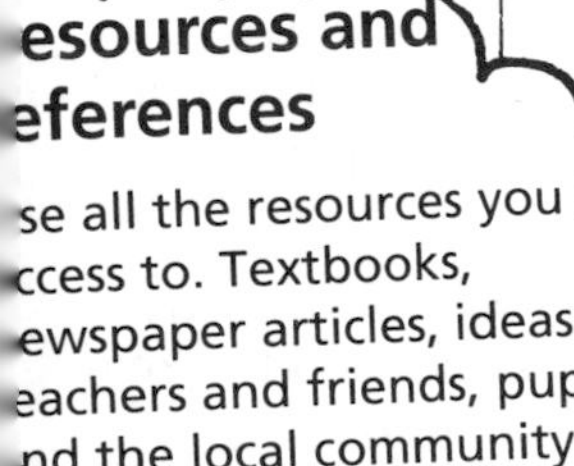

...esources and ...eferences

...se all the resources you have ...ccess to. Textbooks, ...ewspaper articles, ideas from ...eachers and friends, pupils ...nd the local community are all ...ood resources. Exam papers ...re useful for practice ...uestions and to identify which ...opics come up most often in ...he final exam.

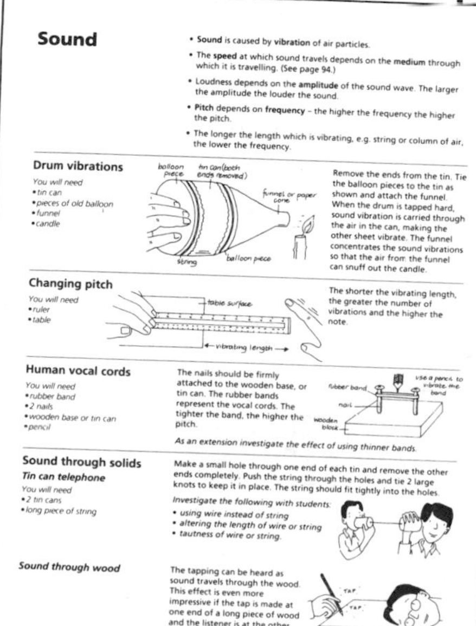

Sound

- **Sound** is caused by **vibration** of air particles.
- The **speed** at which sound travels depends on the **medium** through which it is travelling. (See page 94.)
- Loudness depends on the **amplitude** of the sound wave. The larger the amplitude the louder the sound.
- **Pitch** depends on **frequency** – the higher the frequency the higher the pitch.
- The longer the length which is vibrating, e.g. string or column of air, the lower the frequency.

Drum vibrations

You will need
- *tin can*
- *pieces of old balloon*
- *funnel*
- *candle*

Remove the ends from the tin. Tie the balloon pieces to the tin as shown and attach the funnel. When the drum is tapped hard, sound vibration is carried through the air in the can, making the other sheet vibrate. The funnel concentrates the sound vibrations so that the air from the funnel can snuff out the candle.

Changing pitch

You will need
- *ruler*
- *table*

The shorter the vibrating length, the greater the number of vibrations and the higher the note.

Human vocal cords

You will need
- *rubber band*
- *2 nails*
- *wooden base or tin can*
- *pencil*

The nails should be firmly attached to the wooden base, or tin can. The rubber bands represent the vocal cords. The tighter the band, the higher the pitch.

As an extension investigate the effect of using thinner bands.

Sound through solids

Tin can telephone

You will need
- *2 tin cans*
- *long piece of string*

Make a small hole through one end of each tin and remove the other ends completely. Push the string through the holes and tie 2 large knots to keep it in place. The string should fit tightly into the holes.

Investigate the following with students:
- *using wire instead of string*
- *altering the length of wire or string*
- *tautness of wire or string.*

Sound through wood

The tapping can be heard as sound travels through the wood. This effect is even more impressive if the tap is made at one end of a long piece of wood and the listener is at the other end.

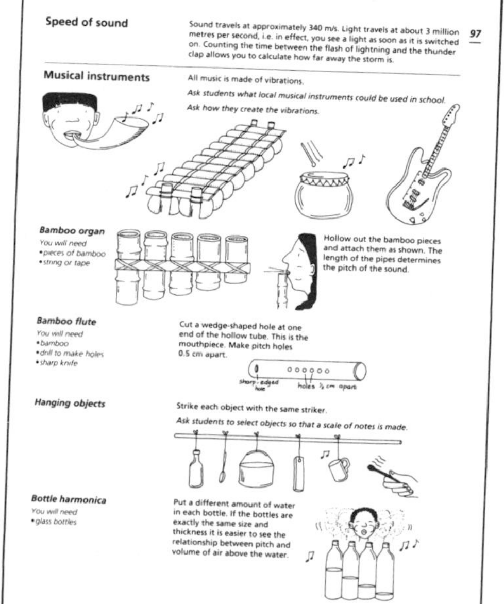

97

Speed of sound

Sound travels at approximately 340 m/s. Light travels at about 3 million metres per second, i.e. in effect, you see a light as soon as it is switched on. Counting the time between the flash of lightning and the thunder clap allows you to calculate how far away the storm is.

Musical instruments

All music is made of vibrations.

Ask students what local musical instruments could be used in school.

Ask how they create the vibrations.

Bamboo organ

You will need
- *pieces of bamboo*
- *string or tape*

Hollow out the bamboo pieces and attach them as shown. The length of the pipes determines the pitch of the sound.

Bamboo flute

You will need
- *bamboo*
- *drill to make holes*
- *sharp knife*

Cut a wedge-shaped hole at one end of the hollow tube. This is the mouthpiece. Make pitch holes 0.5 cm apart.

Hanging objects

Strike each object with the same striker.

Ask students to select objects so that a scale of notes is made.

Bottle harmonica

You will need
- *glass bottles*

Put a different amount of water in each bottle. If the bottles are exactly the same size and thickness it is easier to see the relationship between pitch and volume of air above the water.

Getting the best out of the material

The material in this book should be used in addition to the resources already available, such as textbooks, exam papers, the syllabus, local environment and colleagues. You will need to select items that are relevant to the learning objectives of the students.

The book is divided into 3 main sections: **teaching practice, science ideas, materials and equipment**.

Teaching practice

This section gives information on classroom management and advice on the use of visual aids and other resources.

Try out new ideas first

- It is essential to check new ideas thoroughly in advance to ensure that they work and are safe. If possible work with a colleague and share ideas as well as the equipment with each other.
- If you do not have exactly the materials suggested, try alternatives.
- Trying out an idea yourself may help clarify how it can be adapted for teaching at different levels.

Maximise equipment

If 4 or 5 sets of equipment are available, divide the class into 2 or 3 groups. Each group can then work on an activity for a set time before moving on to the next one. One activity could be a written exercise.

If only 1 or 2 sets of equipment are available consider having a 'circus' of 6–8 different activities that students move around. They could spend only 5–10 minutes at each one. The advantage of such a system is that you make full use of limited resources, but it does require considerable advance preparation and good classroom management.

Use resources fully

Make good use of the resources available. Search the local, national and worldwide community for examples that apply to the topic and make full use of textbooks, newspapers, articles and exam papers.

Take the trouble to find locally based examples of scientific methods and processes in use and also situations where scientific ideas are being applied at industrial level.

Many teachers find it invaluable to meet with colleagues to share ideas and try out new suggestions. Is this possible in your area? Could you become involved in teachers' workshops?

Science ideas

Colour

- **White light** (the light all around us) can be split into its **component** colours by, e.g. a **prism** or water drops.
- White light splits up because its component colours travel at slightly different speeds. Red light bends least, indigo most.
- Filters **absorb**, and so remove, light of particular colours or **wavelengths**.
- The colour of a substance depends on what wavelengths are absorbed by the substance. A red object, for example, looks red because all colours other than red are absorbed by the substance.
- Mixing coloured lights produces different effects to mixing the same colours as **pigments**, e.g. red and green light will produce yellow light.
- **Primary colours** are those which are needed to create all other colours. They are red, yellow and blue, for pigments, but red, green and blue for light.

Breaking up light

Water prism

You will need
- *3 small sheets of glass*
- *adhesive tape*
- *Plasticine*
- *Vaseline*

Stick the 3 pieces of glass together with tape. Use Vaseline along the joints to make them watertight. Push the prism into a base of Plasticine or candle wax so it is watertight. Fill the prism with water.

Investigate how the prism breaks up light into the colours of the rainbow by shining a beam of light through it. *(See page 98.)*

Each spread is a collection of ideas on one topic. The sequence of material is not intended to suggest a progression although linked ideas are grouped together.

Each spread begins with a brief introduction identifying the key concepts being explored.

Where a topic links in with another, or depends upon understanding of another area, cross references are given. In your own scheme of work you could develop many more cross references so that, in effect, you have developed a route through the book.

Materials and equipment

This section gives ideas on sources of chemicals and how to make laboratory equipment from everyday items.

Before making new equipment check that it is worth investing the time and energy required. The criteria for including new ideas given on page 4 may provide a checklist to decide how useful the equipment is. How many criteria can you answer yes to? Which are the most important criteria in your situation?

Another thing to do before you make any equipment is to identify, possibly by a list, the materials that could be used to make the specific equipment you need. If you ask students, friends and shopkeepers to donate things they do not need, you will have a good supply of materials when you want to make something.

Safety goggles

You will need:

- *cardboard, cloth or foam – for padding*
- *glue, Sellotape, masking tape, string*
- *transparent plastic, ideally Melamex*

Make the goggles as shown.

Safety

Some experiments and equipment can be dangerous if not handled properly. Teachers should familiarise themselves with laboratory safety guidelines and take note of safety warnings in the book. Some experiments, especially those flagged by the safety warning shown on the left, may be more appropriate as demonstrations by the teacher.

Particular areas of risk are

- use of chemicals
- heating and cutting glass
- fire
- cross infection, e.g. by sharing apparatus such as straws or blowpipes, or by using unsterilised syringes rather than new or sterile ones.

Extra care is needed when using improvised equipment and all potential risks must be assessed before such equipment is used in an experiment.

While all material has been carefully vetted by experts, neither VSO nor Heinemann Publishers accept liability for accidents of any kind.

Developing new ideas

Do not feel you should rely entirely on your own resources to develop new ideas, involve students and other teachers. Here are some suggestions on how you could involve other people.

Mini-experiments

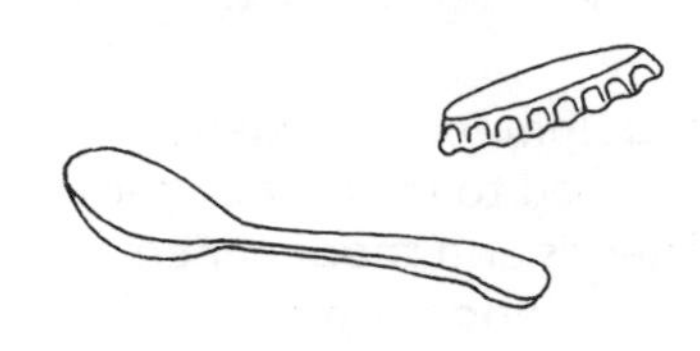

- Smaller scale experiments use up less of your valuable chemicals.
- Spoons and bottletops are not expensive and are easier to replace than specialised equipment.
- Do you really need large vessels?
- Mini-experiments mean more students can carry out experiments themselves.

Involving students

- Students could collect objects for the science department.
- Students could make models or equipment for use in other classes.
- Students could 'act out' a concept or sequence. Some examples are shown below.
- Involve students in evaluating scientific ideas in the context of their own communities. For example, what are their perceptions of science and technology?

Ideas from publications

- Looking through library books and textbooks can give you lots of ideas to use in the classroom.
- Magazines and newspapers may give up to date material.
- There will be other local publications which you could use.
- Encourage students to make full use of their printed resources.

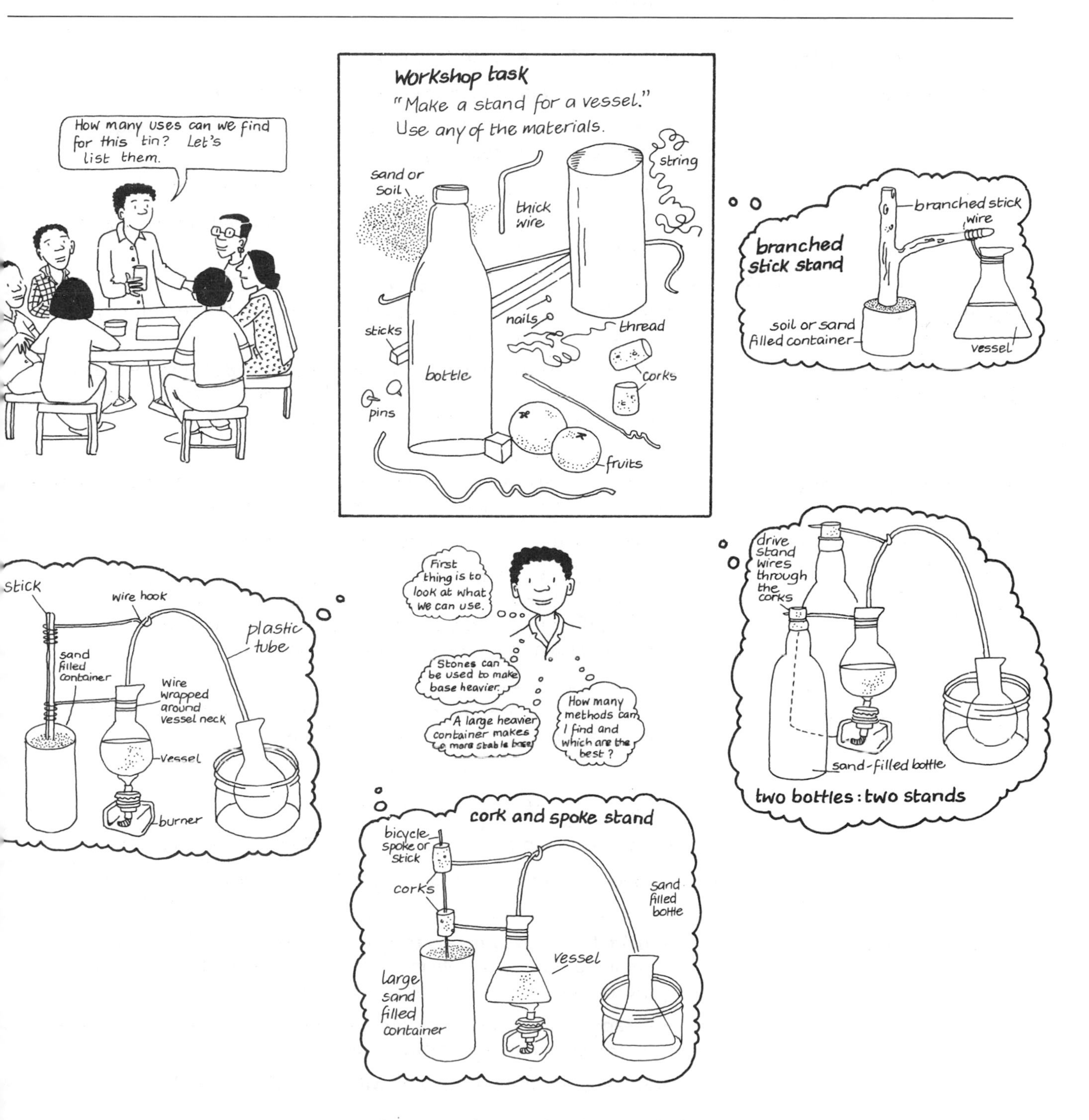

101 uses

Exchanging ideas with other teachers will build up a large selection of ideas for everyone to draw on. It may be useful to have teachers' workshops to develop new ideas and make new equipment. Here is an example of such a workshop project. Students might enjoy the challenge too.

Classroom and community

It is well established that the choice of learning context has a strong effect on student performance. It can even affect the way in which we view our own community. Try to use your local resources to the full and ensure ideas are not presented in a purely theoretical way. Using local examples and local situations to provide illustrations or analogies of science at work will help to bring the subject alive and so motivate students to learn.

New teacher in the community

- Explore the area with teachers who live locally.
- Get to know local technologies, e.g. pottery, bakery.
- Take a notebook and/or camera to record interesting things which may be of use in teaching.
- Get to know the local names for objects, processes etc.
- Discover where materials and plants are situated which may be useful during the year.

New class

- Ask the students to identify what happens in the community which could be described as science.
- A valuable starting point is to ask students to describe their daily routine. When the routine is examined specific events can be identified and developed more easily.

Health and safety

Health and personal hygiene have an important part to play in science education. Here are some examples of starting points.

- Inoculation of a vaccine into the body to give immunity.
- Sleeping in an unventilated room with a wood or charcoal stove may lead to carbon monoxide poisoning. The colourless and odourless gas, carbon monoxide, combines 300 times more readily with haemoglobin than oxygen does. (See page 30)

Science in the home

One of the most useful aspects of the community often ignored in science lessons is the home. Here are some starting points.

- Mixing chemicals, e.g. when preparing foods.
- Stimulation of the production of saliva by smell.
- Using soaps and detergents to break down dirt and fats on clothing.
- Foods going mouldy as a result of bacterial or fungal growth.

Guest speakers

- Invite local craft workers to talk to a class or the science club. Remember these people may not be used to speaking to large groups. Be patient and help if necessary.
- Arrange visits to places of work, where craft workers could explain their jobs in familiar surroundings.
- Build up a register of people willing to talk or visit the school.

Science in the community

Identify local industries which could provide a context for science lessons. Could you develop a science lesson by simulating a local industry in class? Here is an example:

- In many communities plant dyes are made from flowers, roots or fruits. They are collected, crushed and then often boiled for some time before being sieved.
- Batik designs are drawn on cloth using molten wax. The cloth is then dyed, but the dye does not reach the waxed areas. Dyes can also be resisted by tying the fabric tightly.

A science corner

- A table pushed into a corner can be the start of a science corner in the classroom.
- A few nails or strips of wood can be added above the table to hang posters and specimens from.
- The corner could be the focus for science club activities.

Blackboard skills

Presentation

- Untidy presentation encourages untidy work from students.

- If you are right-handed, arrange text on the blackboard so it develops from left to right (vice versa if you are left-handed).
- Divide the blackboard into 2 or 3 fields which are each similar in shape to pupils' books.

- Underline headings and essential terms and statements. Leave space round diagrams. Put summaries in bold or coloured frames.

Cleaning the blackboard

chalk dust everywhere!

chalk dust drawn down

chalk dust pulled to rig

Drawing straight lines

- Use aids to help draw straight lines. Some examples are shown.

a straight-edged piece of wood

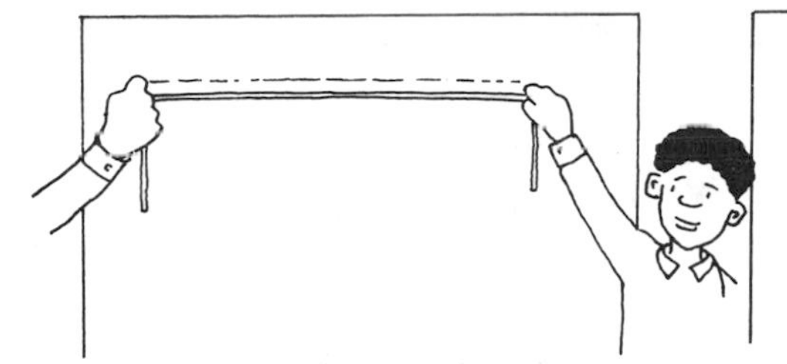

a chalk-covered string... just flick

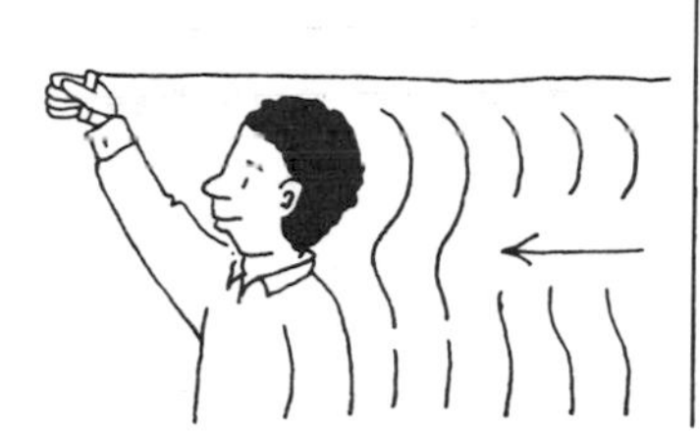

walk a line

Drawing circles

- Use a piece of string, keeping it taut all the time.

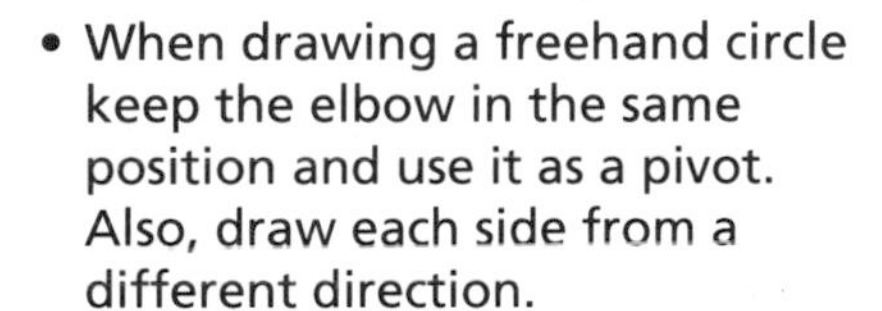

- When drawing a freehand circle keep the elbow in the same position and use it as a pivot. Also, draw each side from a different direction.

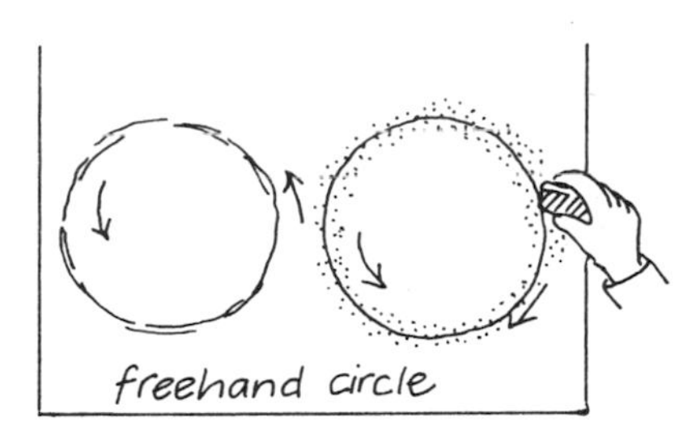

freehand circle

Drawing shapes

- Squares, rectangles and triangles may be produced with knotted string.
- Students can make knotted strings to draw on the board themselves.

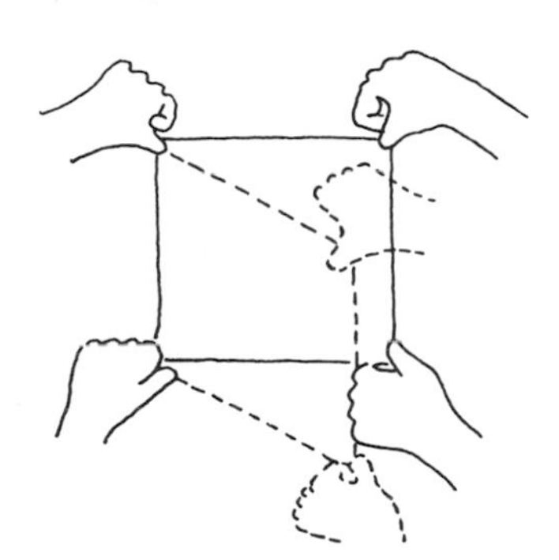

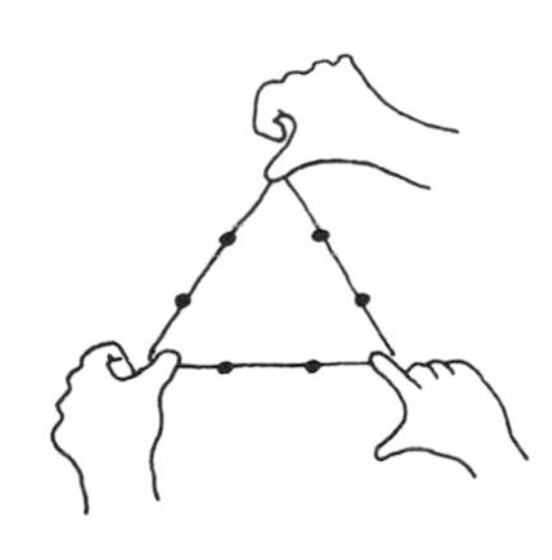

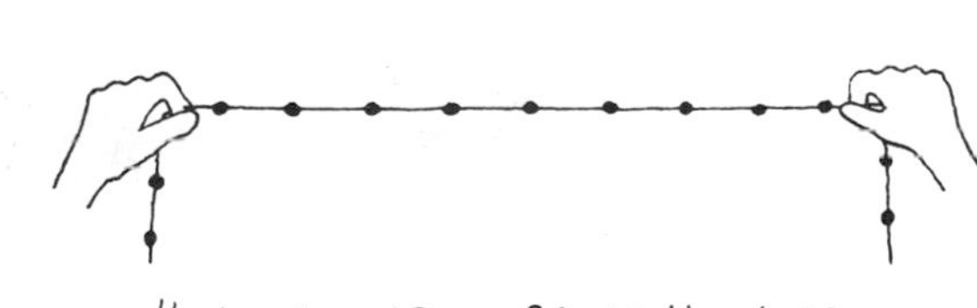

Knot every 10 or 20 centimetres

Different types of line

- Try out different styles.
- A double line may be very effective.
- Experiment yourself.

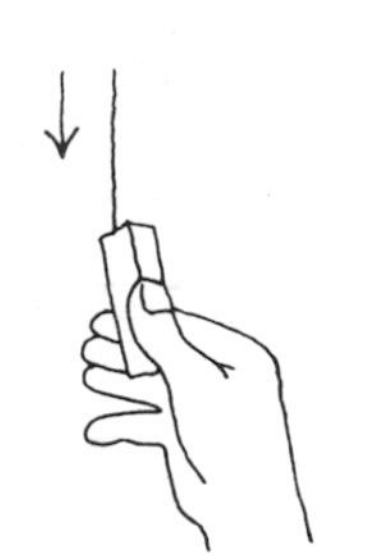

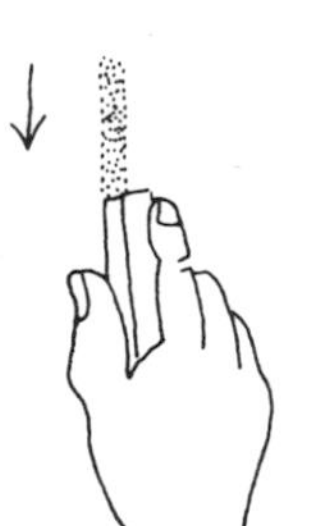

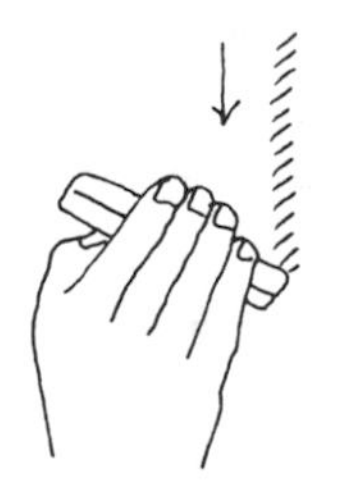

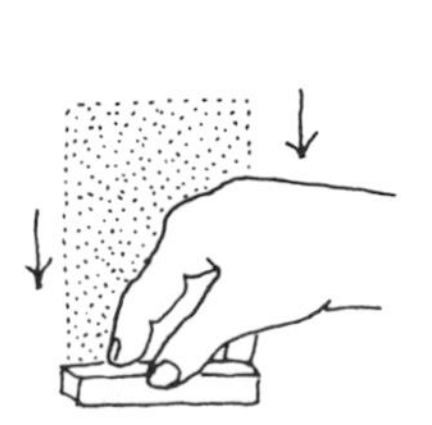

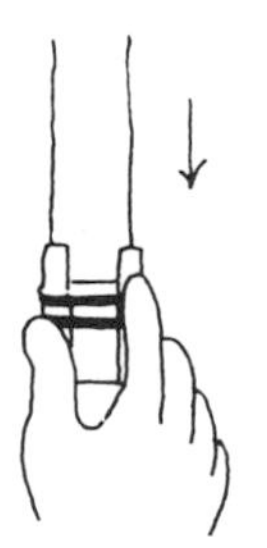

16 Visual aids

Cardboard box display

- Pin display work on the sides of the box.
- Sew or tape cardboard sheets together to make a box (see page 118).
- A box can show 8 sides.

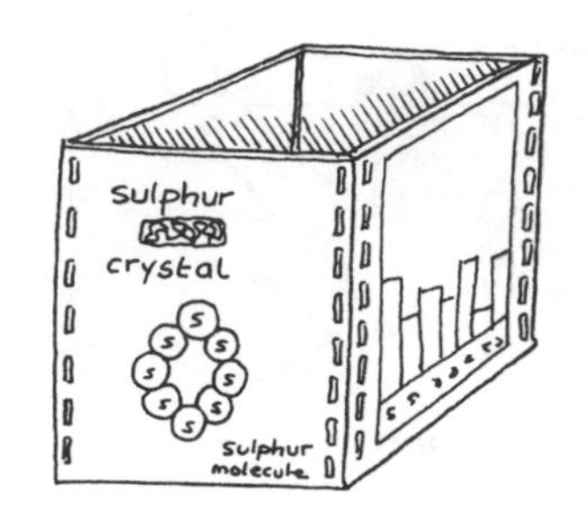

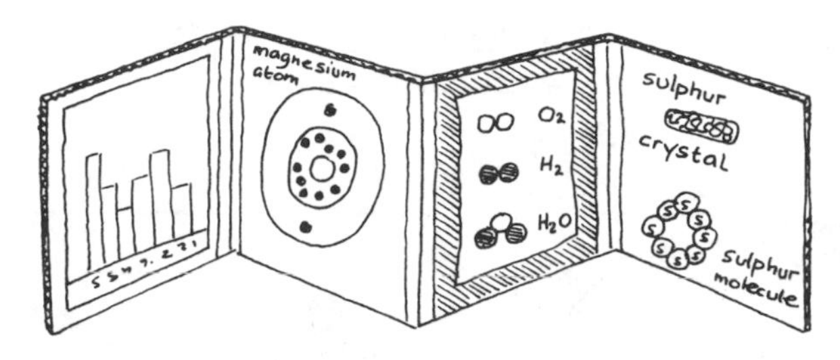

Display beams and hooks

- Make a beam supported by 2 nails or loops of wire. It can be hung on the wall, or suspended from a beam.
- Hooks of wire allow easy and swift display.

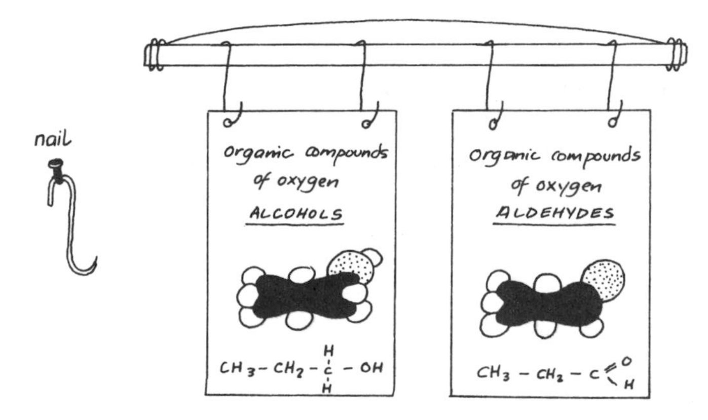

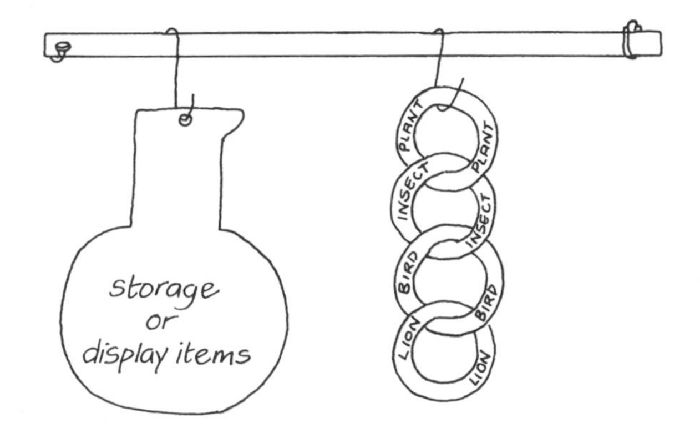

Display charts

- Display charts can be made from durable cement bags, cloth, cardboard boxes, sleeping mats and blankets.
- To make the chart hang flat, attach a strip of wood to the top and either another strip of wood or weights to the bottom.
- Strips at top and bottom will strengthen the chart and make it last longer.
- Attach items to be displayed to the chart with office pins, cactus needles or sharpened matchsticks.

Zigzags

- A portable zigzag board can hold and display many items.

Zigzag multiboard

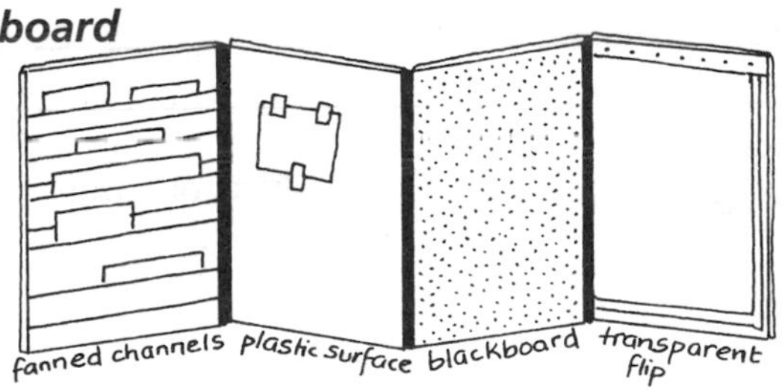

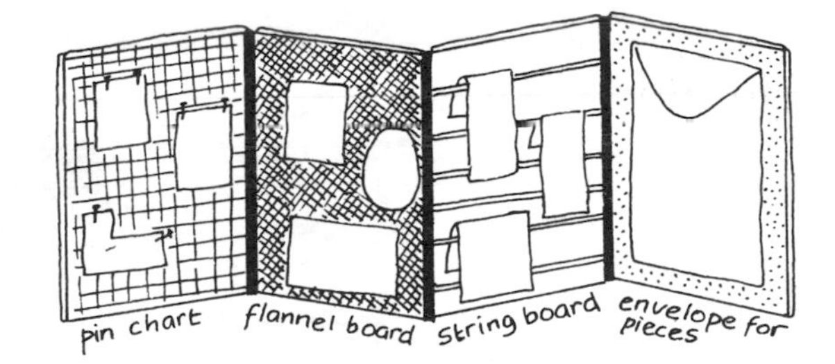

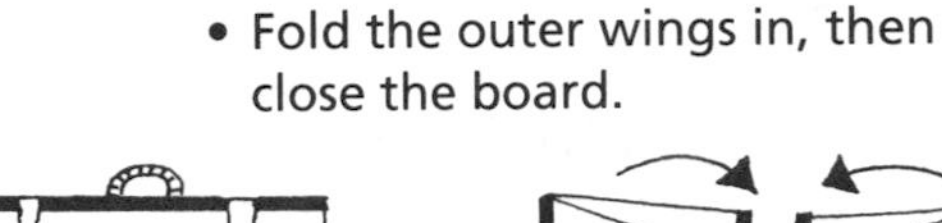

Zigzag board portability

- Fold the outer wings in, then close the board.
- The boards can be made from plywood, hardwood or cardboard.
- Fastenings can be made from many materials.

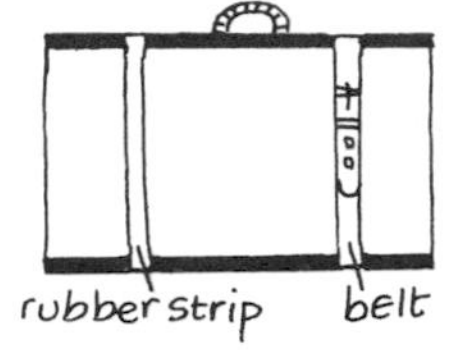

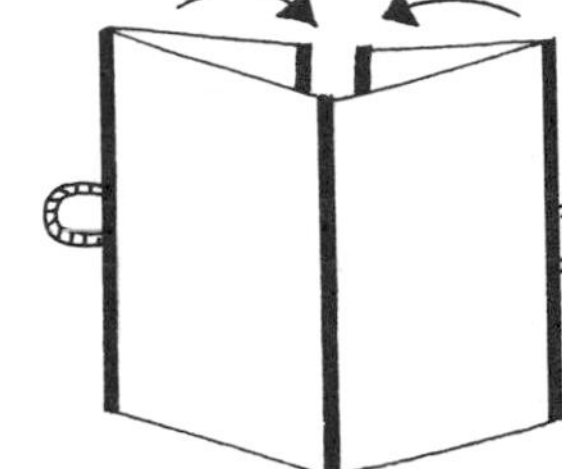

Zigzag variations

- Experiment with different angles and presentation techniques.
- Try different combinations in one board.

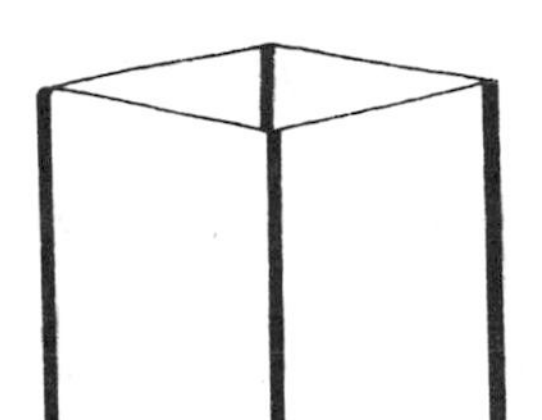

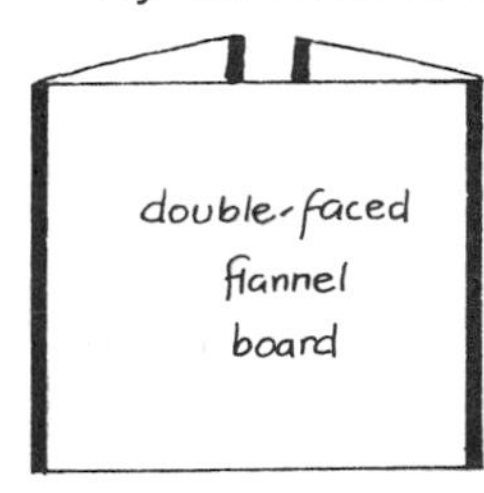

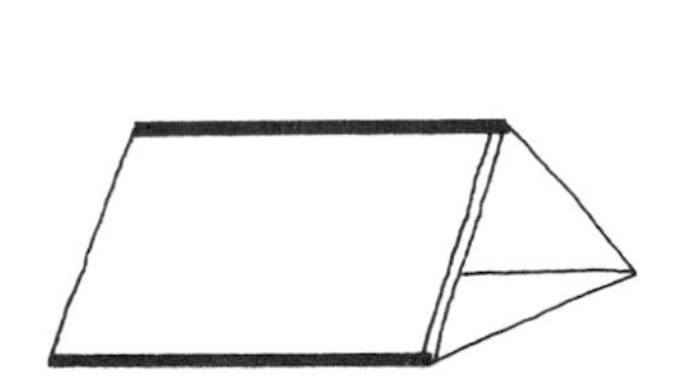

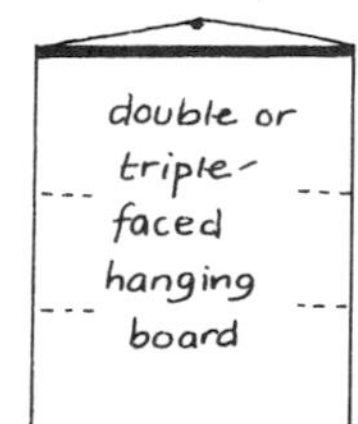

Flannel/cloth boards

- To keep pictures firmly on a cloth or flannel board, attach a small piece of sandpaper to the reverse side of the picture. Press the picture onto the board.
- Alternatively, sprinkle sand on a lightly glued area on the back of the picture to give a sandpaper effect.

See-through flip sheets

- You will need polythene sheets, a bar of wood and some nails or pins.
- You can put together as many sheets as you want (see page 33).
- Lift up different sheets to show the combinations you want.

1. draw the base drawing

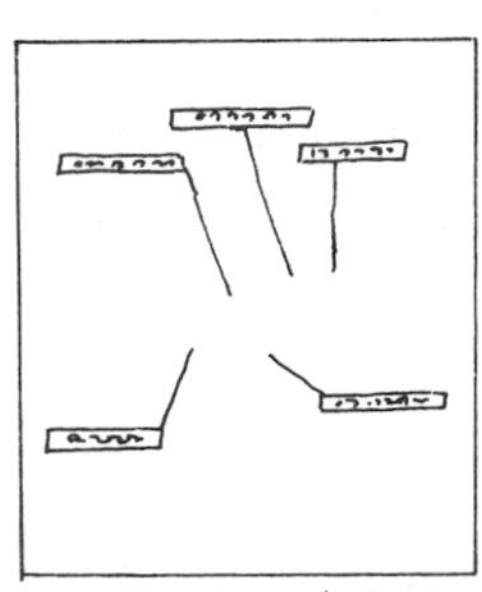

2. write the labels on the other sheet

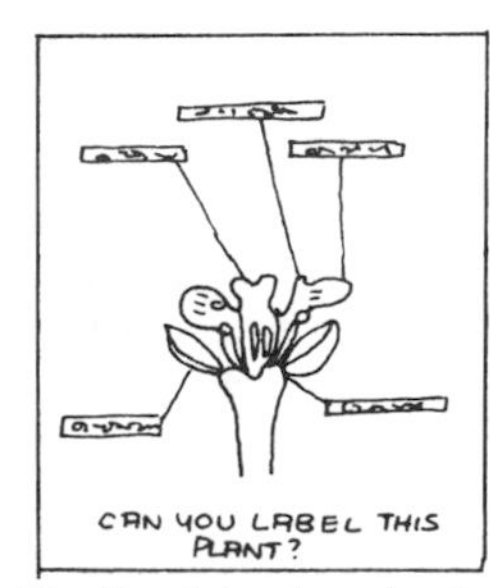

3. the two drawings lying one over the other

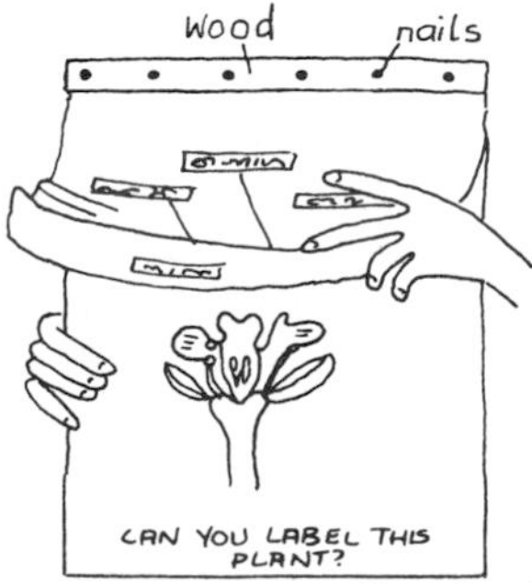

4. using the flip chart

Visual aids *continued*

Clothing posters

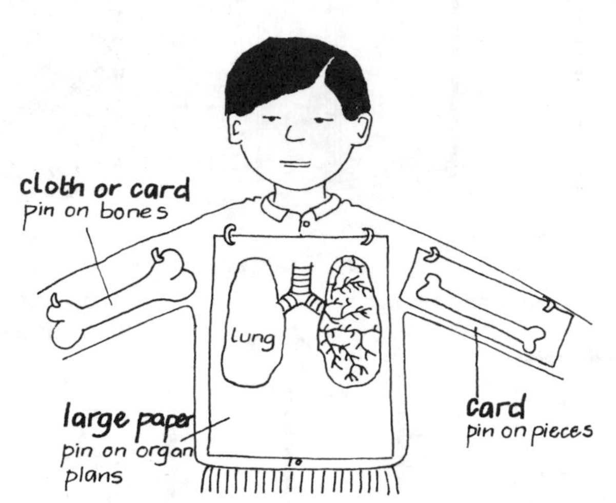

Body organs could be drawn, painted or pinned onto gloves, T-shirts or trousers.

Magnet board

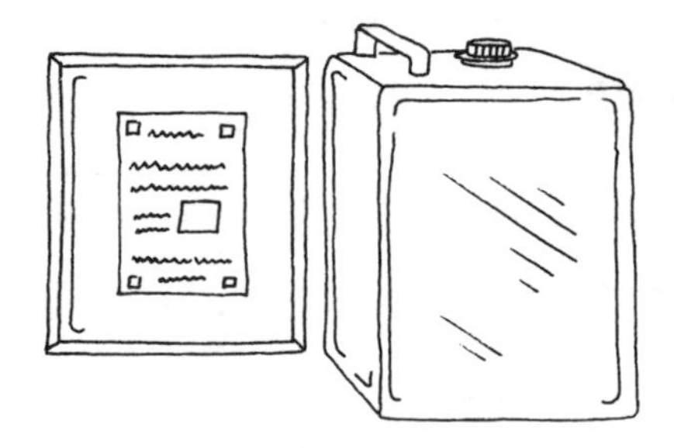

- Use a thin metal sheet. Paint it black to act as a blackboard too.
- Metal could come from old cans or car panels, fridge doors, filing cabinets, steel shelves, flattened corrugated sheet.
- Tape over the edges of the sheet, or hammer the edges over for safety.
- Magnetise small pieces of metal (see page 74) to attach pictures to the metal sheet.
- Painting the metal pieces white makes them less noticeable. Glue the magnetic pieces to the back of pictures used regularly.

Hanging displays

Carrier bag display

- Open out a plastic carrier bag and tidy the edges.
- Attaching a wooden stick at the top and bottom of the carrier bag adds strength and makes it hang flat.
- Permanent marker pens can be used to draw onto the plastic (some come off with spirit). Alternatively use washable markers.
- Use Sellotape tabs to attach pieces to the display chart. These can be movable pieces.
- Paper attached by flour paste (see page 118) can be washed off with water.

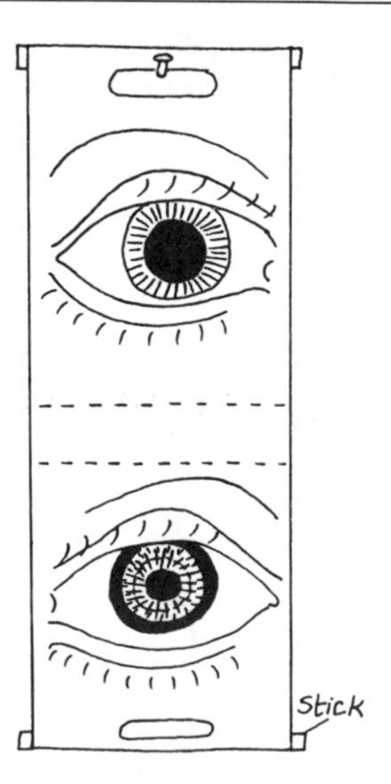

Hanging mats

- Hanging mats can be made from heavy cloth, sleeping mats or even carpet.
- Attach sticks to the top and bottom to give extra strength.
- Attach pictures or posters with tailor or office pins, sharpened matchsticks or palm frond vanes ('broom pieces').

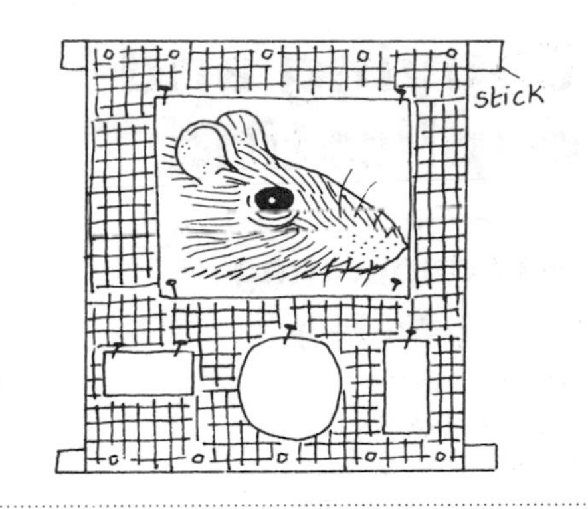

String display lines

- String can be used in many ways to display items. Some ideas are given here.

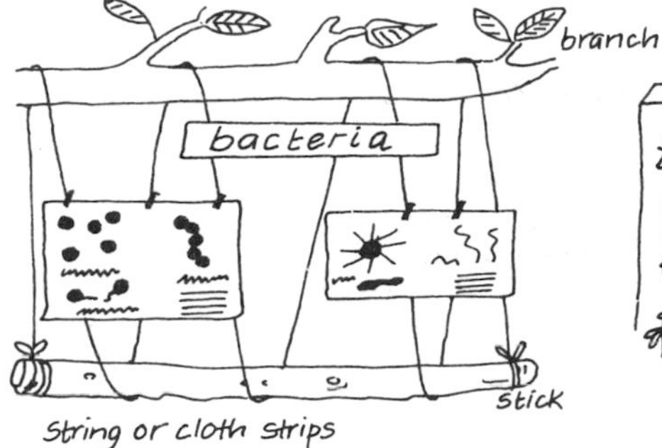

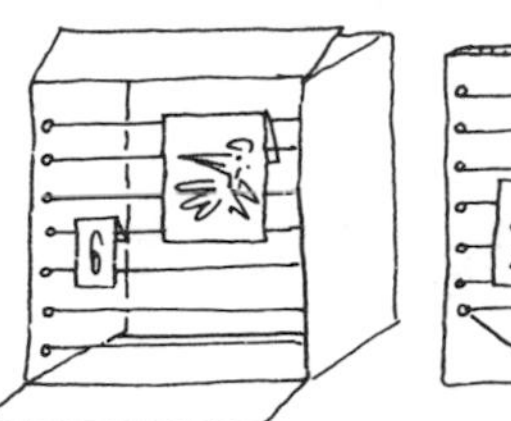

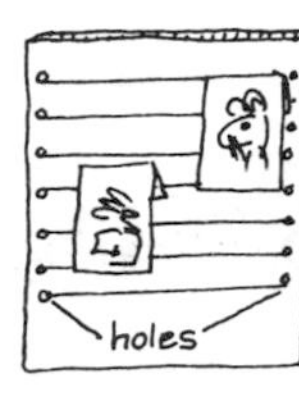

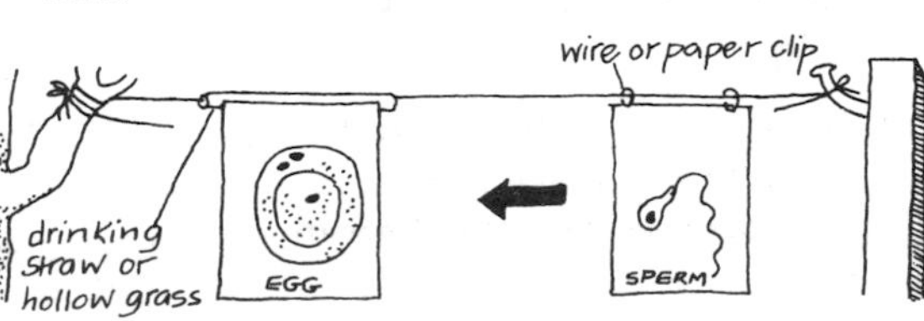

- Hollow tubes, e.g. drinking straws, or paper clips will allow the display to slide up and down the string.

Templates and stencils

Blackboard and book templates

- Templates can be cut from hardboard, plywood or cardboard.
- Collect basic and elaborate shapes for quick, uniform and accurate reproduction.
- Put a hole in blackboard templates so they can be hung up for storage.

Stencils

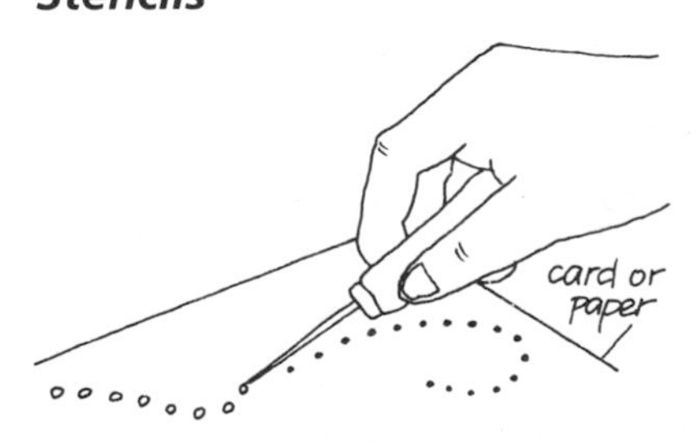

- Draw the shape onto cardboard then punch guideline holes through the cardboard, following the outline of the drawing.
- Pin or hold the stencil against the blackboard or wall.
- Dust over the holes with chalk.
- Remove the stencil and simply join the dots to reproduce maps, charts and diagrams on a large scale.

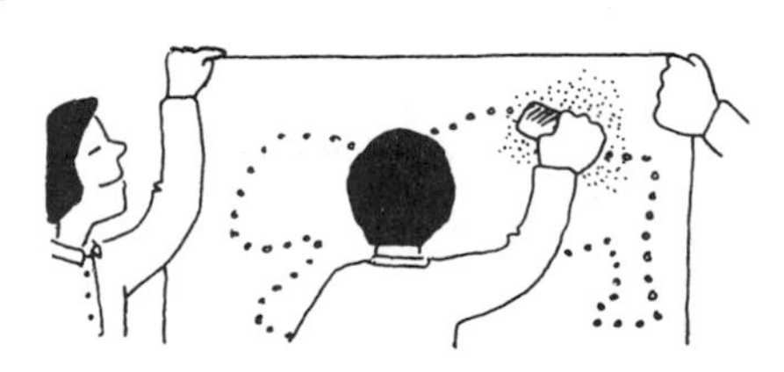

Copying and duplicating

Jelly duplicating

Making the jelly

You will need

- *1 packet gelatin (11 g)*
- *1 cup boiling water*
- *2 teaspoons sugar*
- *50 ml glycerin/glycerol*

- This recipe should be adapted to local conditions. Try less water in hotter climates to help the jelly set and more sugar to act as a preservative.
- Work quickly through all the stages.
- Mix all the ingredients into the boiling water and stir well until they have dissolved.
- Pour the liquid into a shallow tray large enough to take A4 paper.
- Remove any bubbles.
- Leave the liquid to cool and set in the tray.

Duplicating

- Prepare the original as shown.
- Wet the jelly surface slightly with a sponge.
- Bend one corner of the original down so you have a 'flap' to hold it.
- Place the original face down on the jelly and leave it for 2–3 minutes then remove it from the jelly.
- Bend one corner of the copy paper (to make a lifting flap) and place the paper on the ink-impregnated jelly.
- Remove the copy paper after a few seconds.
- It is possible to duplicate 20–30 times.

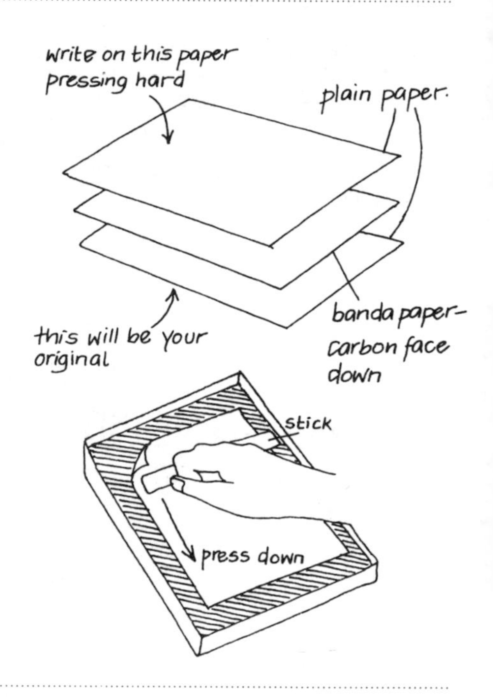

Jelly copier hints

- Allow the impregnated jelly to sit for 12 hours before re-use. The ink then soaks in and the jelly can be re-used.
- Melt a used jelly down and the ink diffuses throughout the jelly. The jelly can then be re-used. Melt jelly over water, never with a direct flame.
- Make stencils (like a Banda original) on any paper except newsprint.
- Smooth, glossy, strong paper is good even if printed on.
- Purple and red pens often work best, but this may not be true for all types of inks.
- Try commercially produced jelly too.

Making tracing paper

- Brush or dab a little turpentine onto a sheet of white paper.
- Work swiftly when doing this because turpentine evaporates quickly.

T-shirt and cloth prints

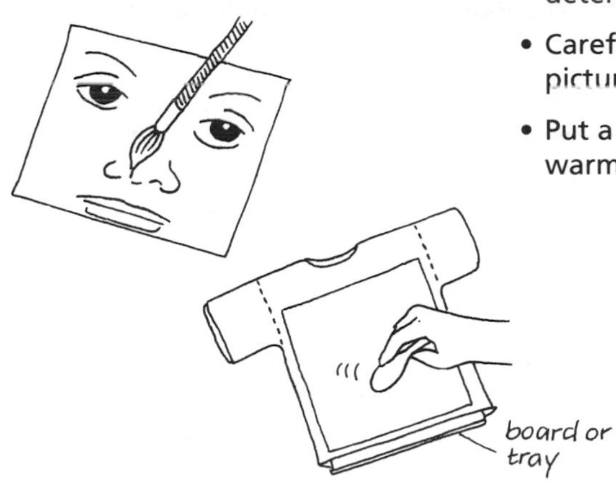

- Mix 1 part white spirit with 2 parts water and a little soap or detergent.
- Carefully brush the mixture onto the front and back of a photocopied picture.
- Put a flat board under the cloth surface you want to print onto. A warm metal tray helps transfer best. Try foil on the board instead.

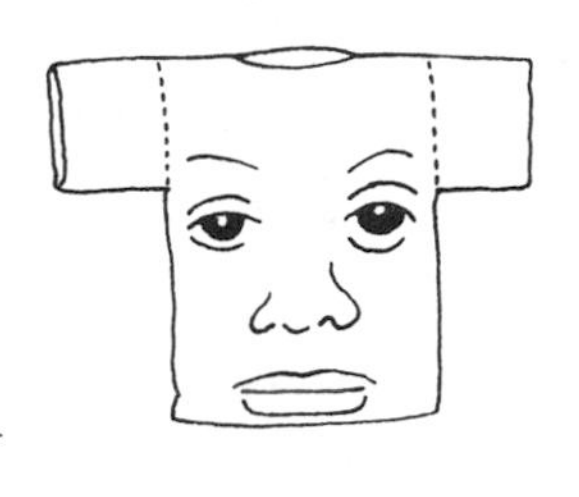

- Use the bowl of a spoon to press the back of the photocopy onto the cloth.
- Spray or paint on art fixative or camping waterproof fluid for permanence.

Reverse transfers

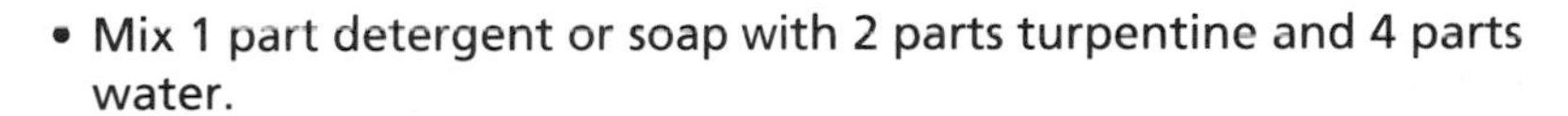

- Mix 1 part detergent or soap with 2 parts turpentine and 4 parts water.

- Shake the mixture until it forms a white emulsion.

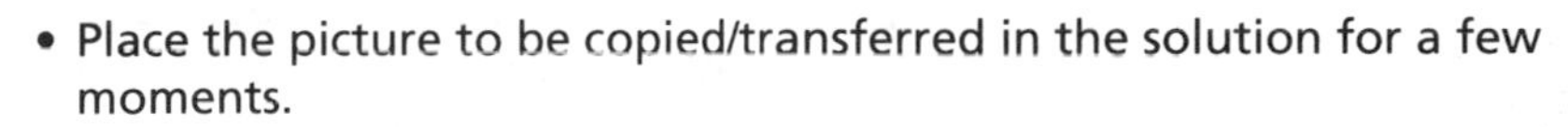

- Place the picture to be copied/transferred in the solution for a few moments.

- Remove the picture and place it face down on a new piece of paper.

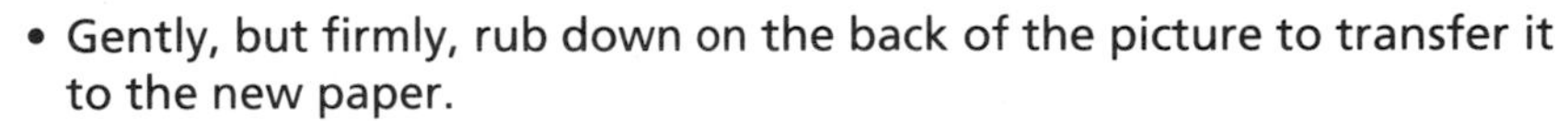

- Gently, but firmly, rub down on the back of the picture to transfer it to the new paper.

Paint or ink pad printing

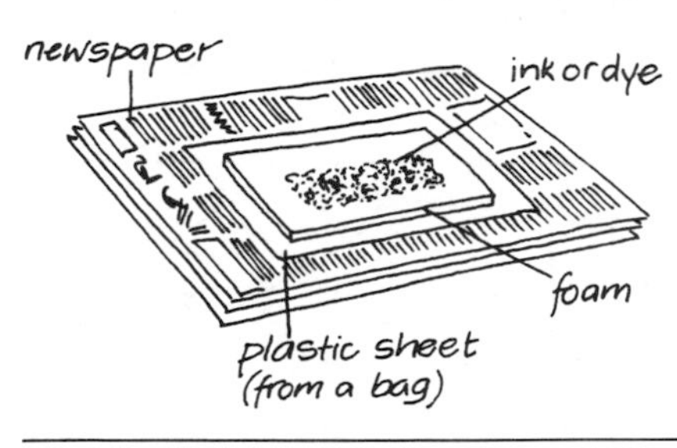

- Make the ink pad as shown.
- The printing block can be cut from various substances, e.g. potato, soft wood, thick sheet rubber. Try other surfaces to print from.

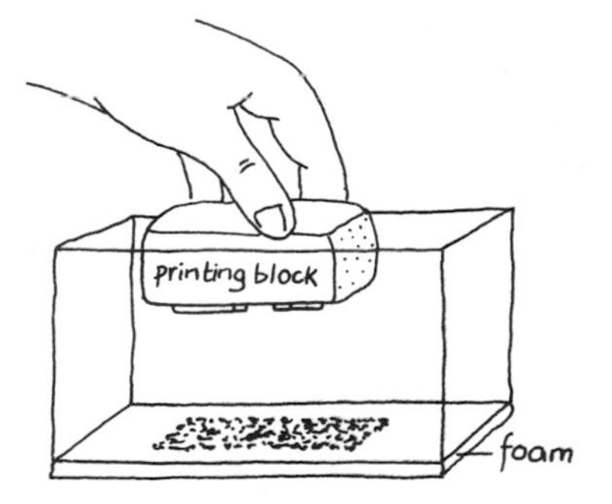

Making poster paints

- Mix 1 heaped tablespoon of corn starch with a little water to form a paste.
- Slowly add 1 cup of boiling water to the paste, stirring constantly to prevent lumps forming.
- Simmer the mixture until thick and smooth.
- Add 1 tablespoon of detergent or soap.
- Add dye (see page 101).
- Experiment with this method to find the best results for your materials.

Cells and tissues

- All living things, except **viruses** and **bacteria**, are made up of cells.
- All cells have a **nucleus**, which contains the **genetic material** (see page 52), and surrounding **cytoplasm**.
- The main difference between plant cells and animal cells is that plant cells usually contain **chloroplasts** (see page 38) and cell walls (see hydrostatic skeleton, page 40).
- The membrane surrounding the nucleus and the **cell membrane** are both **semi-permeable** (see page 24).
- **Cells** join together to form **tissues**, e.g. kidney tissue or skin.
- Tissues join together to form **organs**, e.g. kidney and heart.

Cells, tissues, organs

You will need
- *matchboxes*
- *peas/beans/stones*
- *boxes of different colour or size*

Place a seed in each box. This represents the nucleus: the matchbox the cell. Place groups of cells inside the coloured boxes – the different coloured boxes represent different tissues and the boxes themselves can be joined to make organs.

The school is a useful model of an organism. The bricks (cells) make walls (tissues) and walls make classrooms (organs). The corridors can therefore be used as models for transport systems (see blood, page 30). Another analogy might be a town where buildings represent organs, rooms the tissues or cells and people inside the rooms the various functions of the cell.

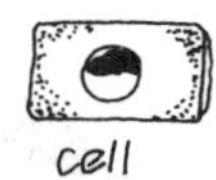

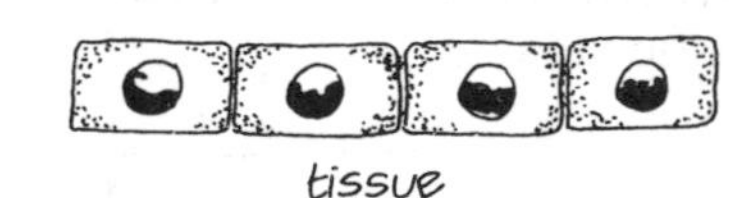

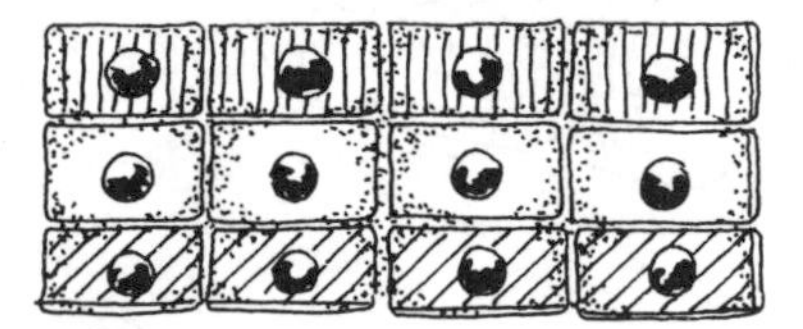

Models of cells

You will need
- *2 large and 2 small plastic bags*
- *water*
- *2 large seeds or stones*
- *small seeds or coloured paper*
- *grass*
- *cardboard box*

Make models of plant and animal cells as shown. This idea can be adapted for primary and A-level students.

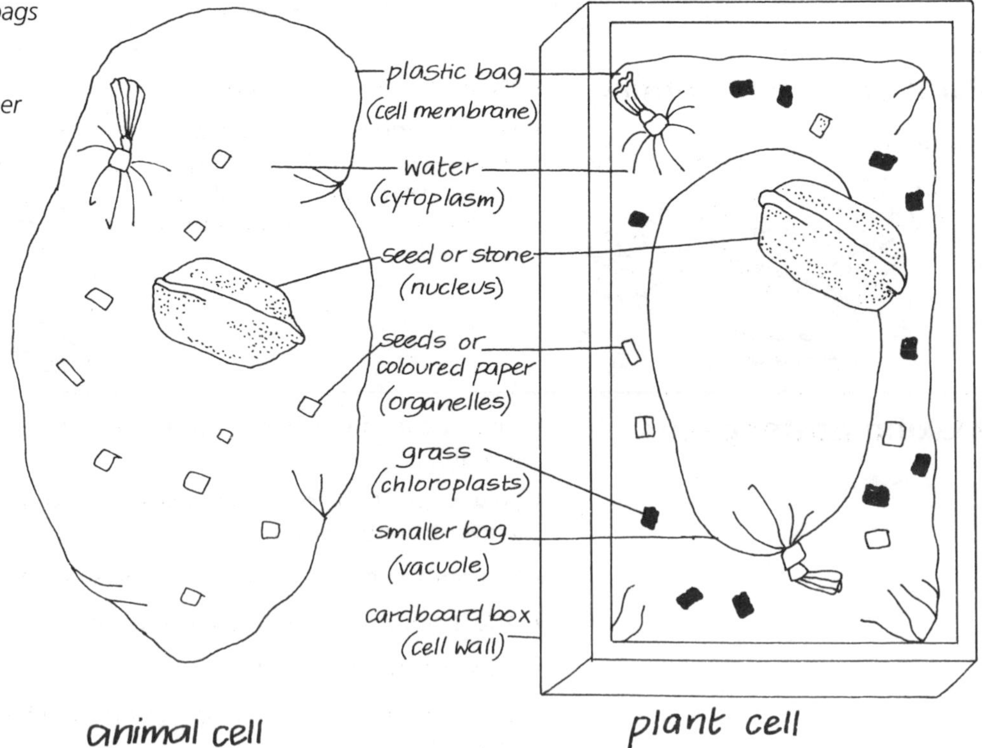

Simple home-made microscope

You will need

- *soft drinks can*
- *small lens e.g. from a pen-torch bulb (see page 115)*
- *aluminium strip*
- *small mirror*
- *piece of glass*
- *rubber band*

Make the microscope as shown. Some care is needed in positioning the lens in the hole made for it in the aluminium strip. The inside of the can may be painted black. Such a microscope is quite adequate for looking at cells.

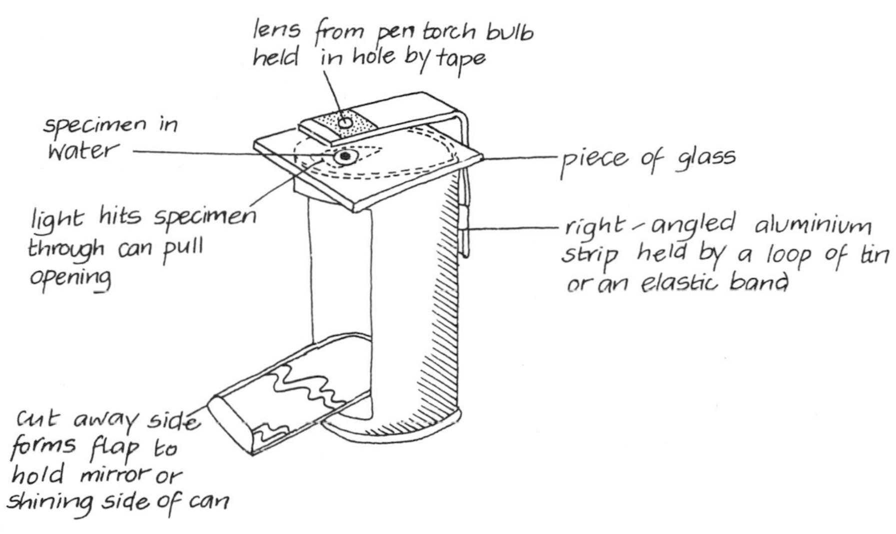

Looking at cells

You will need

- *an onion*
- *pin or needle*
- *glass*
- *cover slip*
- *tweezers (see page 110)*
- *iodine solution*

Cells from an onion bulb are particularly easy to see, but other similar bulbs may be used instead.

Cut a slice of onion and gently peel off a piece of the thin inner surface skin layer. With a pin/needle place a piece of 'skin' in a water drop on a piece of glass. Stain the 'skin' with a drop of iodine solution. Lower a cover slip onto the specimen taking care not to let in any air bubbles. (The thin plastic used in display packaging is very good for cover-slips.) Now view the prepared slide through the microscope (see cell structures page 22).

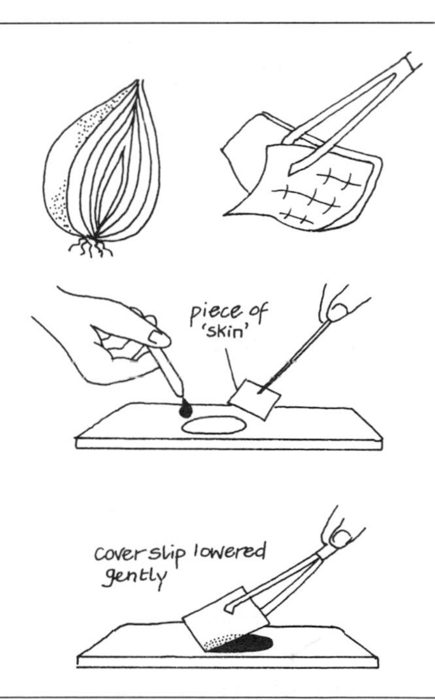

How many cells in a human?

This activity is a useful way of explaining to students how small and numerous cells are. If you use sand to represent cells, you need to point out that each grain is several thousand times larger than a human cell.

Ask students to estimate how many grains of sand would be needed to make a model of a person. They could make estimates by counting how many spoonfuls are needed to make a human and then estimating the grains in a spoonful.

Ask students to make a dot with a sharp pencil and point out that the largest human cell, the ovum, is smaller than this.

24 Diffusion and osmosis

- **Diffusion** is the process by which molecules move from an area of **high concentration** to one of **low concentration**.
- A **semi-permeable membrane** allows only small molecules to pass through it.
- **Osmosis** is the net flow of water across a semi-permeable membrane from the area of low concentration of the solute to that of higher concentration of the solute.
- Osmosis can be thought of as a special case of diffusion because, although all molecules may 'want' to move until their concentrations either side of the membrane are equal, only the small water molecules can pass through the semi-permeable membrane.

Diffusion – a model

You will need
- *glass of clean water*
- *coloured ink*
- *pipette or straw*

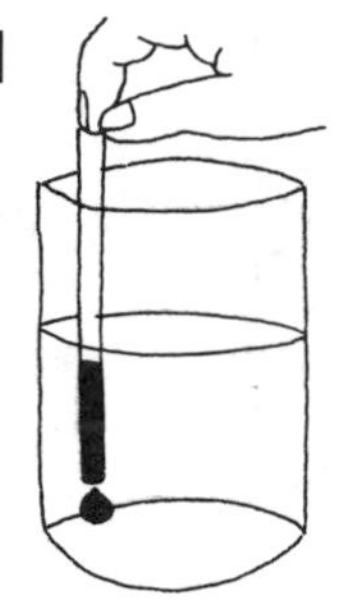
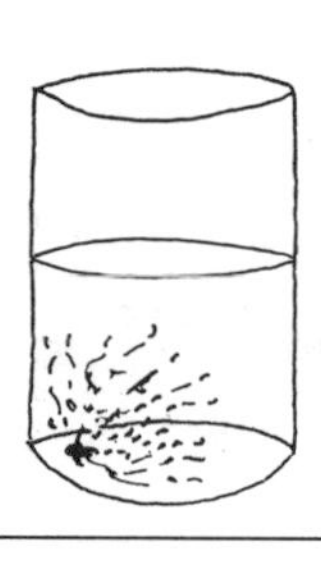
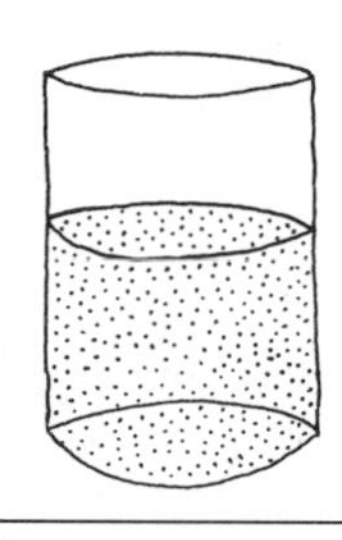

A drop of ink placed carefully on the bottom of a glass full of water will diffuse into the water until the colour (and the concentration of ink) is evenly distributed.

Semi-permeable membranes

Model shaker

You will need
- *glass jar*
- *transparent plastic bag*
- *small beads or stones*
- *larger peas or beans*
- *netting*

Place the mixture of beads and beans in the jar. Place the net and the plastic bag over the top and tie them on securely. When you shake the apparatus only the beads (small molecules) pass through the net (semi-permeable membrane).

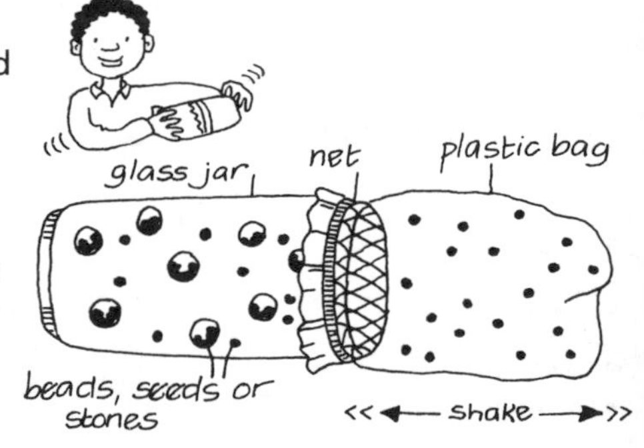

Membrane analogy

A net bag is a good analogy of a semi-permeable membrane. The small objects will fall through the holes, but the larger ones remain – the net is semi-permeable.

Demonstrating a semi-permeable membrane

You will need
- *glass container*
- *pipette or straw*
- *clear plastic bag filled with corn starch and water*
- *iodine solution*

Some, but not all, plastic bags are semi-permeable, you will need to check the suitability of your bag. Toy balloons are semi-permeable.

You will notice that eventually the iodine stains the corn starch blue/black, indicating it has crossed the membrane. The water in the container remains clear however, so the starch has not crossed the membrane.

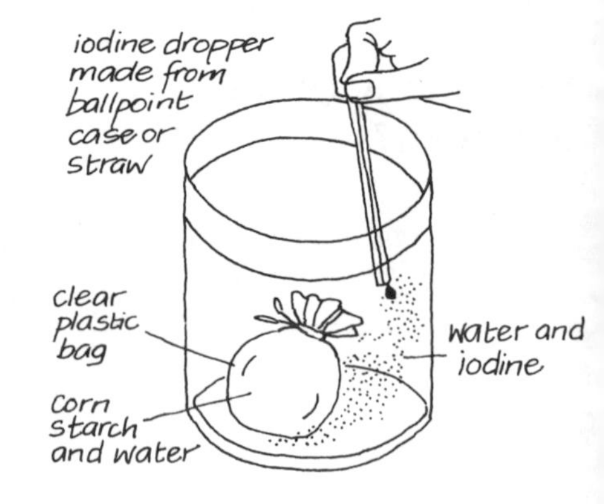

Osmosis

Osmosis with eggs

You will need
- *empty eggshell*
- *strong salt solution*
- *jar containing water*

Remove the hard outer shell at one end of the eggshell. This will expose the inner membrane. Half fill the egg with salt solution and place it in the jar so that the water level is above the exposed membrane and leave for a couple of hours. You will see the level of the solution inside the egg rises, indicating water has crossed the membrane, i.e. osmosis has occured.

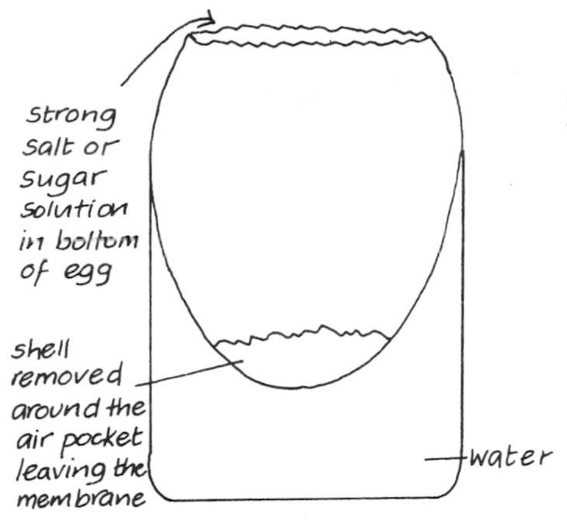

Ask students to use sugar solution instead of salt and discuss their results.

Ask what happens if they put salt solution in the jar as well as the egg.

Investigate with students what happens if the salt solutions inside and outside the egg are of different concentrations.

Osmosis in dead and living tissues

You will need
- *yam, paw paw or potato*
- *strong sugar solution*
- *dish of water*

Cut the yam or paw paw in half and boil one piece. When it has cooled, hollow out the centre of both pieces of yam and half fill the hollows with the sugar solution. Peel the lower half of both yams and then place the pieces in a dish of water for an hour or so. Water will only enter the unboiled yam, if it enters both you need to boil the yam longer.

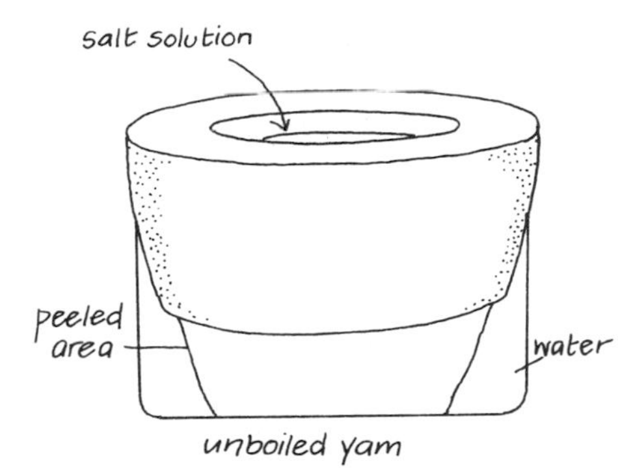

Investigate with students how boiling affects the semi-permeable membranes of the yam.

Ask students which yam has living cells in it and also to explain what happens if they do not peel the yams.

Guard cells – osmosis in practice

You will need
- *2 stretched long balloons*
- *thick adhesive or rubber tape*
- *rubber band*

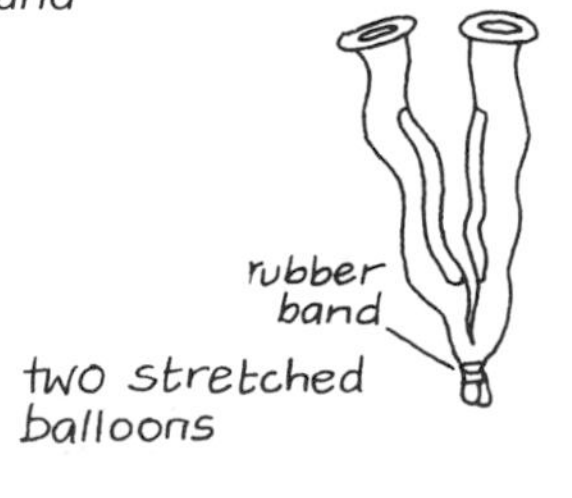

Guard cells become turgid or flaccid due to changes in osmotic pressure. This can be seen with a microscope or demonstrated using balloons.

Stick the adhesive tape down one side of each balloon as shown. When the balloons are both fully inflated (turgid) the 'stoma' is open. If you let out some of the air, i.e. the 'guard cells' become flaccid, the 'stoma' closes.

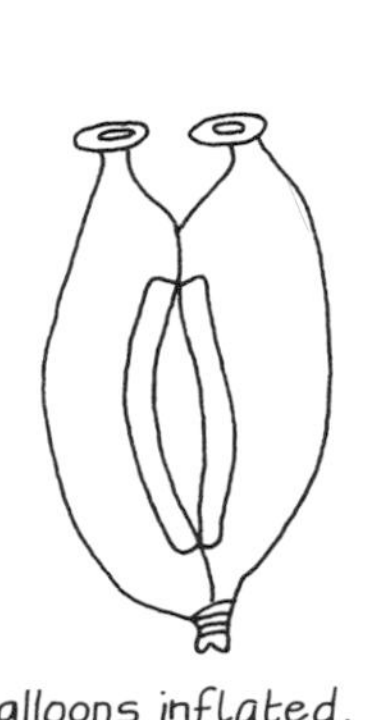

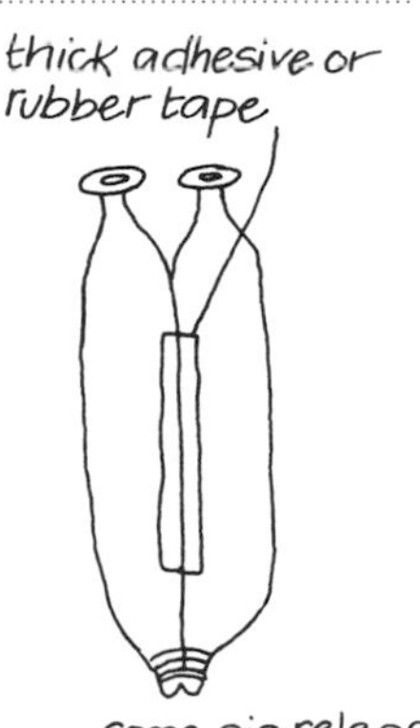

Osmosis in the kitchen

You will need
- *dried seeds or fruit to soak*
- *suitable container*
- *water*
- *salt solution*

When dried fruits or seeds such as figs, apricots, peas or beans are soaked in water they swell up as water is absorbed by osmosis.

Ask students to see whether the same thing happens with salt solution.

Ask whether the concentration of the salt solution affects the swelling.

Ask why cooks often put salt onto sliced vegetables before cooking.

26 Foods and food tests

- Foods are divided into 3 groups: **carbohydrates, fats** and **proteins**.
- A healthy human diet requires a **balanced diet**.
- Many foods are rich in only 1 or 2 food types, e.g. bread is almost entirely the carbohydrate starch.
- During **digestion** the large **molecules** are broken down into their smaller component molecules by **enzymes**.
- Only small molecules can pass through the **semi-permeable** wall of the alimentary canal (see semi-permeable membranes page 24).

Sorting foods

You will need
- *a selection of foods or pictures of foods*

Ask students to sort the foods into the three main groups. Then ask them to identify the main use of each type, e.g. carbohydrates and fats for energy, protein for growth etc.

Ask students to research the effects of insufficient minerals or vitamins.

Ask students to group foods by function, e.g. high energy foods, foods which require little digestion.

Fats and oils – the lipids

You will need
- *card*
- *scissors*

Fats are made up of glycerol and fatty acids.

Cut out the shapes of the glycerol and fatty acid molecules. They can be combined to form fat (lipid) molecules. The longer the fatty acid chains the more solid the lipid. Oils have short chains of fatty acids, fats much longer ones.

Ask students to form fats of different types with the cards.

This could be made into a game.

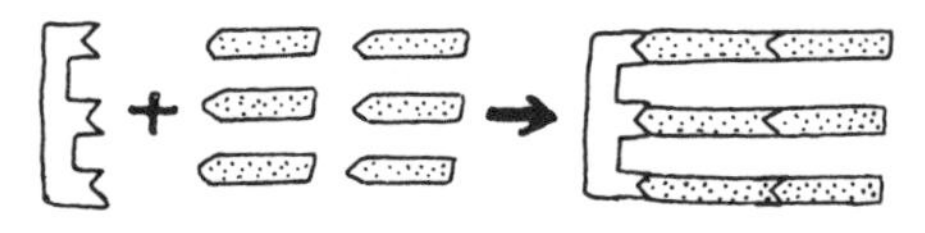

Carbohydrates

You will need
- *peas, beads or any set of identical items*

Each pea is a monosaccharide, e.g. glucose. Putting 2 together makes a disaccharide, e.g. table sugar and a long chain of them a polysaccharide, e.g. starch. Toilet roll provides another analogy of the way identical units combine in long chains to make a polysaccharide. *Note:* Not all di- and polysaccharides consist of identical units, e.g. sucrose is a disaccharide of 2 monosaccharides glucose and fructose.

Models of food molecules

You will need
- *bottletops*
- *seeds*
- *beans*
- *fruits*
- *paper or card*
- *string*
- *scissors*

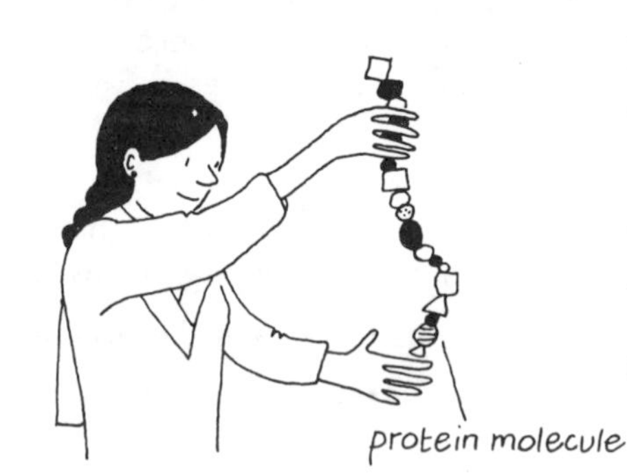

A variety of different shaped and sized items threaded on a string show how different types of amino acids join together to make a protein molecule. Students can collect their own materials and make their own models, or they could cut out shapes from paper or card.

Models for digestion

Starch is a polysaccharide made up of many identical glucose molecules. Proteins are made up from many different amino acids. During digestion large molecules are broken down into smaller ones by enzymes, e.g. starch is broken down into glucose, proteins into the component amino acids. You can cut up models of food molecules to demonstrate digestion.

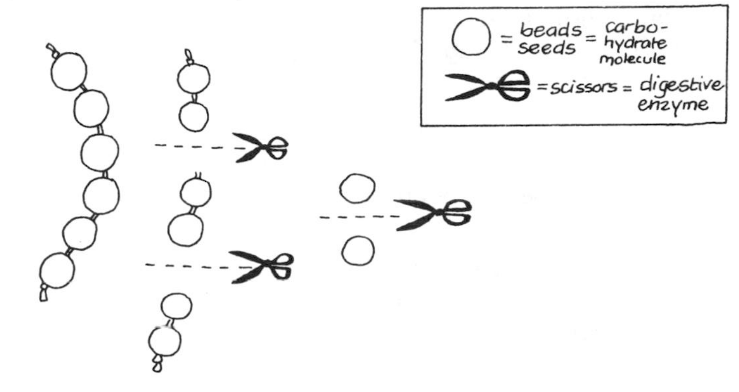

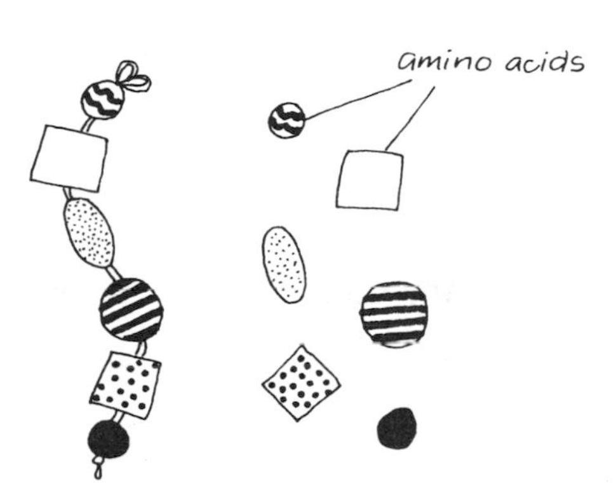

Food tests

Each of the 3 food groups can be identified by a simple, standard food test. If the food is not in liquid form crush a little up in a small amount of water for use as your test substance. See page 105 for instructions on how to prepare the chemicals used here.

Protein
Add sodium hydroxide to the test substance then add copper(II) sulphate solution to it. Purple colour confirms the presence of protein.

Starch
Add iodine solution. Blue-black colour confirms the presence of starch.

Reducing sugars
Dissolve the food in water. Put some into the bottle top and add Benedict's or Fehling's solution. Heat very gently for 1 minute. Safety goggles should be worn. If a precipitate develops – usually green or brown – this confirms the presence of sugar.

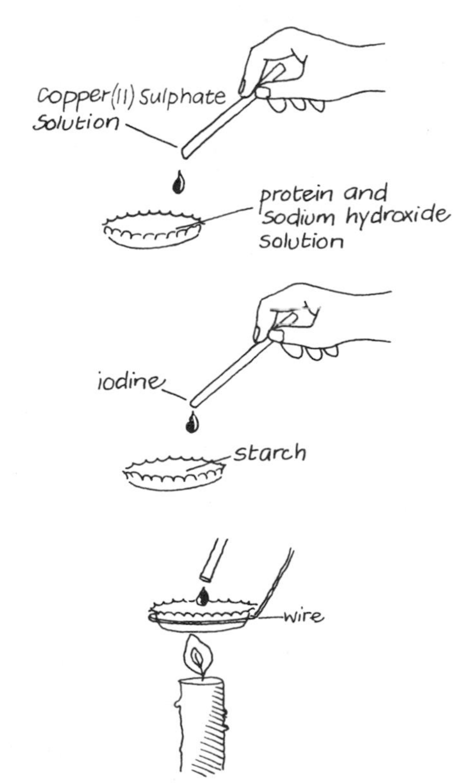

Fats and oils
Rub a piece of food onto a piece of paper. Fat is present if there is a translucent stain.

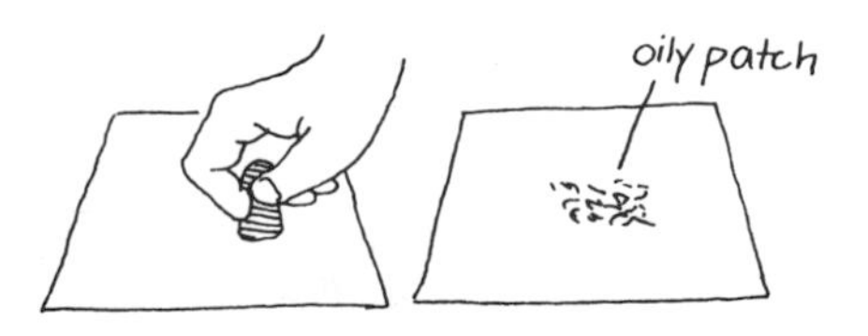

Alimentary canal and digestion

- The **alimentary canal**, or **gut**, is a long tube starting at the **mouth** and ending at the **anus**.
- **Involuntary muscles** control the movement of food along the gut by **peristalsis**.
- Glands secret **enzymes** into the gut which **digest** the food as it moves along towards the anus.
- The gut is a **semi-permeable membrane** (see page 24) and small digested molecules pass through it, i.e. they are **absorbed** into the blood vessels surrounding the gut.
- Once in the blood, small molecules are carried to various parts of the body and small molecules may be combined, **synthesised**, to form large molecules again (see page 26).

Alimentary canal

A model pig

You will need
- *transparent plastic bottle*
- *thin tubing*
- *card*
- *plastic beaker*

Make a model animal as shown – the model may be adapted to depict a different animal if a pig is not a suitable choice. Make sure the tube sticks through the bottom of the bottle.

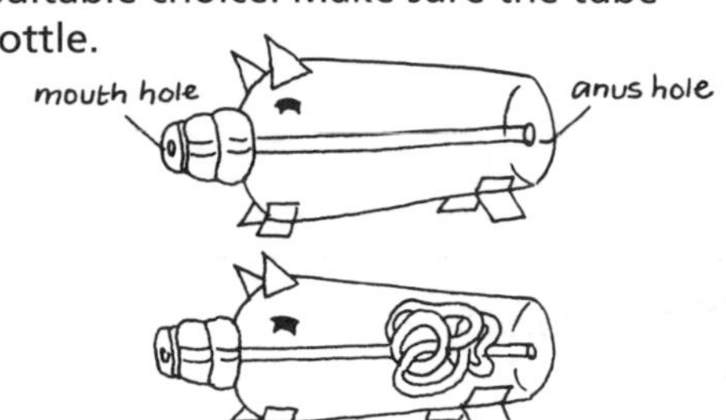

Improve the model by making small holes in the tube to show absorption.

Discuss with students how long it takes water to pass down the model's tube.

Intestine length

You will need
- *long piece of rope or paper strip*

Draw the shapes of different animals on the ground. Use string, rope or paper strip to show the length of intestine and coil it into position on the animal. Approximate lengths are: rabbit 1 m, cat/dog 2–5 m, horse 30 m, cow 50 m, human 5 m.

Ask students why intestine lengths differ and why herbivores have longer intestines than carnivores.

Model of digestive system

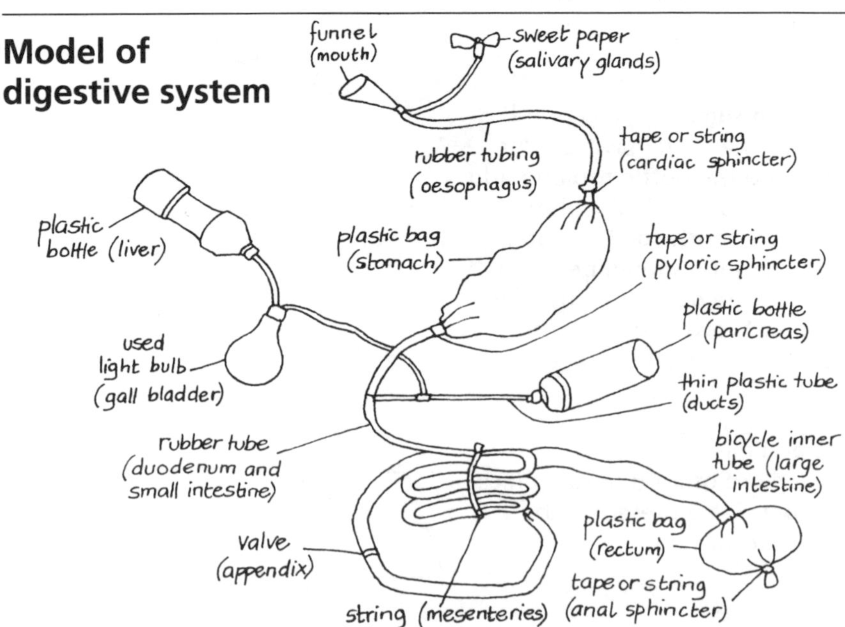

You can make a model of a digestive system using the everyday objects shown.

Extend the activity by colouring different sections and labelling them then mounting the model on a display board (see page 16).

Ask students to place the model inside a box so it demonstrates how the intestine passes through the diaphragm.

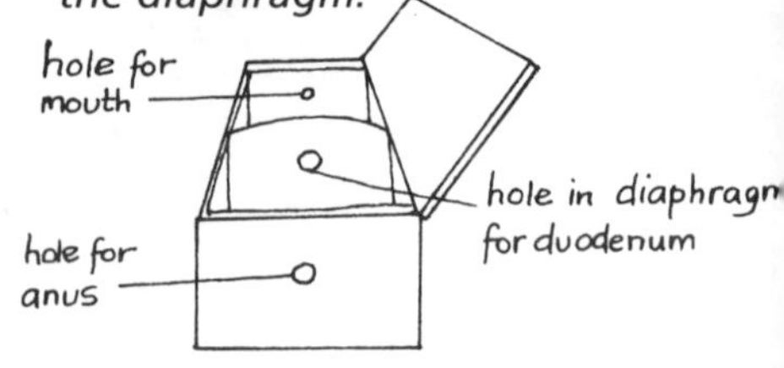

A model of peristalsis

You will need
- *a balloon*
- *rubber tubing*
- *ball or seed*

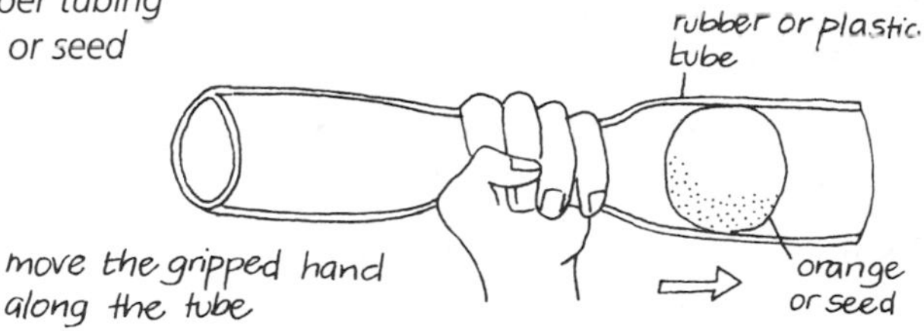

Food is moved by the contraction of the muscular walls of the gut. A balloon gripped with the hand pushes air along. You can also move an object along a tube by squeezing behind the 'food' ball.

Ask students how else they could make models, e.g. using a bicycle inner tube.

A model of absorption

You will need
- *an old shirt sleeve*
- *small objects, e.g. peas*

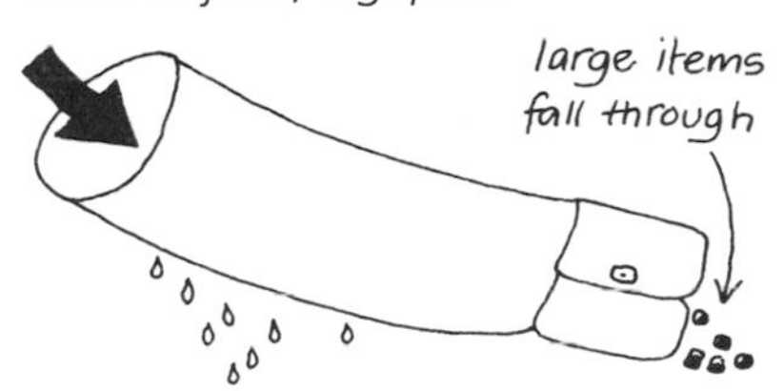

Place the shirt sleeve over a container to catch the water as it drips through. Pour the mixture of water and peas down the tube. Water will leak out, but the peas (undigested food) pass straight down. You may need to tie off the end of the sleeve to slow the process down.

Ask students to improve the model, e.g. by using several layers of newspaper instead of a shirt sleeve. (Not all ideas will work well, so test them.)

Extend the activity by using a semi-permeable plastic bag for the gut. Pour starch and sugar into the tube and test to see what passed through (see page 26).

Digestion of starch

Chewing

You will need
- *different types of food*

Ask students to chew different types of food for a long time before swallowing. They should notice that starchy foods seem to get sweeter as the saliva digests the starch to sugars.

Ask students why starchy foods taste sweet after long chewing.

Enzyme action

You will need
- *filter paper*
- *matchsticks*
- *starch solution*
- *iodine solution*

***Safety*: Make sure students do not share matchsticks, doing so could pass on infections.**

Soak filter paper in starch solution. Ask students to use saliva on a matchstick to write their names on the treated paper. Dip the filter paper in dilute iodine solution.

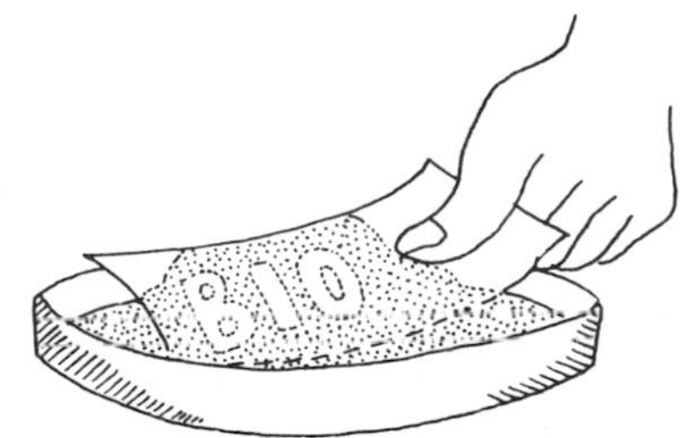

Ask students what causes the name to appear on the paper.

Enzymes – protease activity

You will need
- *strips of exposed film*
- *pineapple or paw paw juice*
- *gelatin*

Dip the strips of film in a fresh solution of pineapple juice or pounded paw paw. You will see that the solution breaks down the gelatin coating containing the black silver salts and the plastic is left clear.

Ask students the following questions.

What happens to a piece of gelatin left in the juices?

What happens if you pour the pineapple or paw paw juice over boiled egg white or meat and leave it for some time?

Why is paw paw called a meat tenderiser? What does it do to meat?

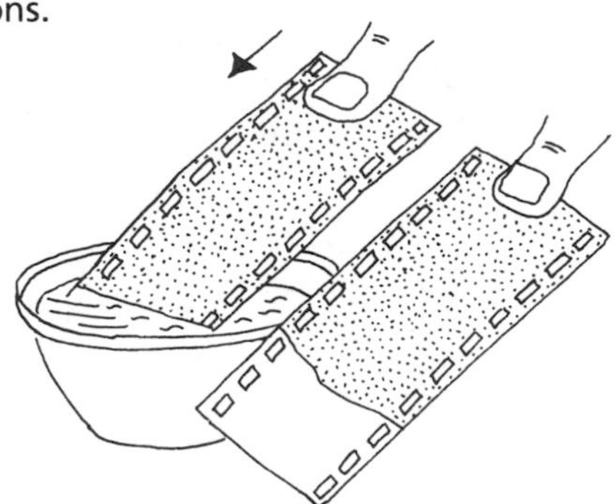

Blood

- The adult human body contains about 4–5 litres of blood.
- Blood **circulates** round the body carrying with it **dissolved** food (e.g. after digestion) and **waste products** (e.g. to the kidneys).
- **Red blood cells** contain **haemoglobin** which transports oxygen from the lungs to the **tissues**.
- **White blood cells** are an essential part of our **immune system** because they destroy **bacteria** by engulfing them.
- In humans there are several different **blood groups**.

Blood the transporter

Blood brings substances to the cells, e.g. food and oxygen, and removes others (waste and CO_2). A food bar or shop has items delivered, gives out items and produces waste. This gives a good analogy for the blood system. Students can act out the role of blood by picking up or putting down items at different shops (sites of the body).

Ask students what they pick up and put down at the following sites: lungs, liver, muscles, kidneys…?

Models of red and white blood cells

You will need
- *Plasticine, clay or wooden rod*
- *card or sponge*

Red blood cells are biconcave discs with no nucleus. You can make models from Plasticine or circles of wood. White blood cells could be cut from thin sponge rubber sheet. They contain a nucleus which can be drawn in on the sponge. Platelets, essential for clotting at open wounds, can be made from smaller, irregular pieces of sponge, clay etc.

Make red and white blood cells by cutting shapes from cardboard, paper or plastic.

Add platelets and then put everything into water. Ask students what the water represents.

Discuss the functions of each part of the blood.

Engulfing model

You will need

- *clear water-filled plastic bag or cloth*
- *stone or bean*

Partly fill a clear plastic bag with water. Put a stone or bean inside to represent the nucleus. By shaping the bag, the action of a white blood cell engulfing a foreign body can be demonstrated. You could use a cloth, handkerchief or blanket as a white blood cell. Shape the cloth to show the pseudopodia surrounding the foreign body.

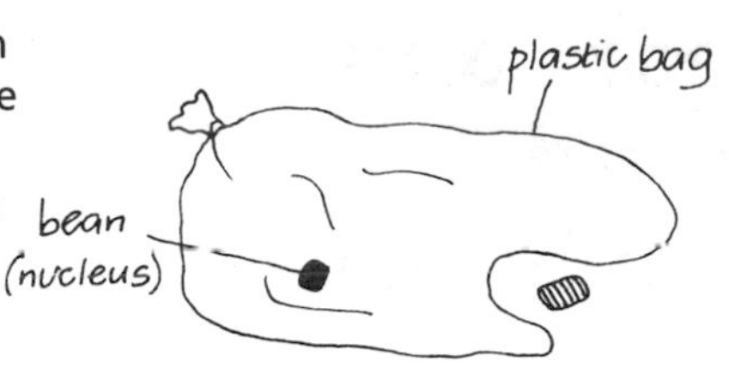

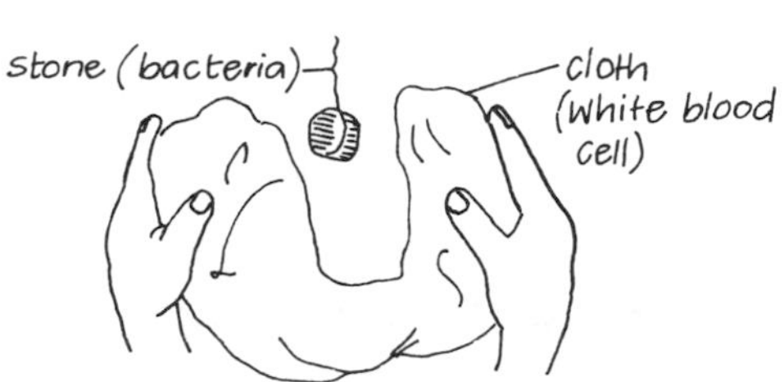

Transfusion games

Transfusion checkers

You will need

- *2 types of bottletops or counters*
- *card*
- *coloured pens*

Draw out a base grid as shown. Use 2 types of bottletop or counter to show 'safe' or 'clot' transfusion.

Ask students to place the tops on the correct squares.

Ask students if they can think of an easy way to remember the formation.

Ask students to identify which blood groups are compatible.

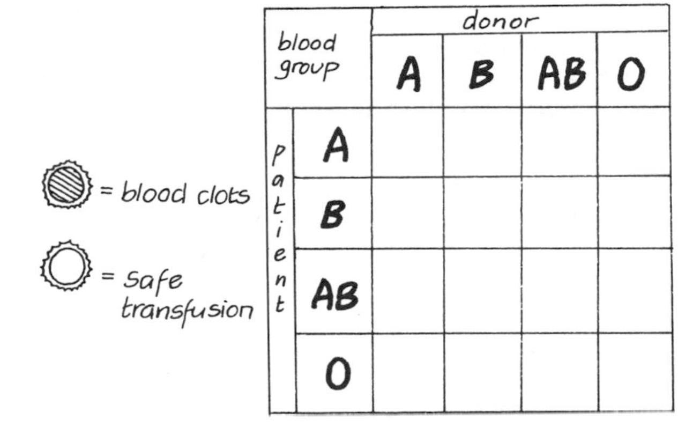

blood group	donor A	donor B	donor AB	donor O
patient A				
patient B				
patient AB				
patient O				

blood group	donor A	donor B	donor AB	donor O
patient A	○	●	●	○
patient B	●	○	●	○
patient AB	○	○	○	○
patient O	●	●	●	○

Transfusion card game

You will need

- *card*

Cut out 20 cards and label 5 for each blood group. The dealer shuffles the pack face down on the table and then turns up one card face up. This is the patient's blood group. The next card turned over is the donor's blood group. If a transfusion is possible, players must call 'safe'. If a transfusion would be dangerous they call 'clot'. The first player to call correctly wins the 2 cards. The player with the most cards wins the game.

32 Heart and blood circulation

- **Blood** fills the small tubes – **veins, arteries** and tiny **capillaries** – which form the **blood system**.
- Blood is **circulated** by the pumping action of the **heart**.
- **Valves** in the veins and heart ensure blood circulates in only one direction.
- During exercise the **heart beat** increases and the blood circulates more quickly.
- We can detect the heart beat either by listening to the heart itself, or by feeling the **pulse**.

Measuring the pulse

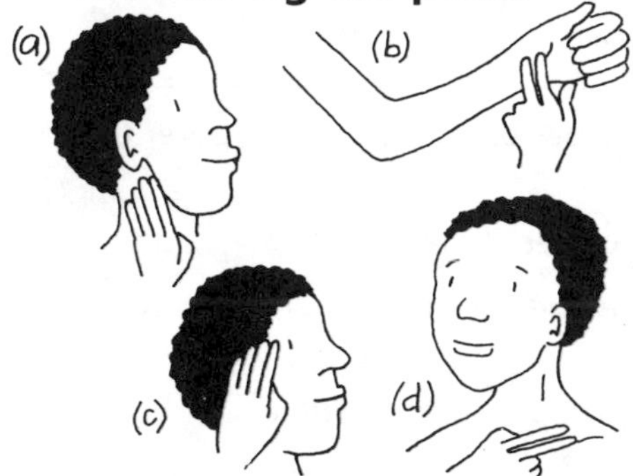

Students can listen to their own pulse beat. Putting their fingers in their ears cuts out external noise. The muffled rhythmic pulse beat can then be heard.

There are various places on the body where the pulse may be taken. They are (a) under the ear beside the angle of the jaw, (b) at the wrists, (c) at the temple, (d) behind the collar bone.

Ask students to find the pulse of a partner. If they have difficulty, they should move their fingers around or apply a little more pressure.

Students could compare their partner's pulse rate before and after exercise.

Making a stethoscope

You will need
- *newspaper or the top of a plastic bottle or a funnel and rubber tube*

A stethoscope focuses the sound of the heart. You can make one by using the materials shown. Place the stethoscope against the ribs or back and then listen to the heartbeat.

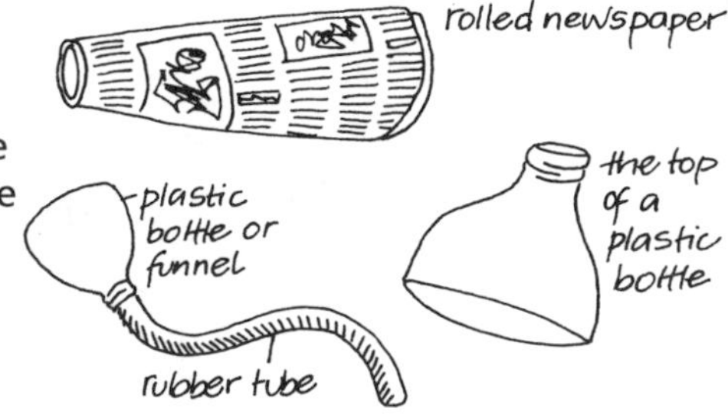

The heart

Heart pump action

You will need
- *2 buckets or bowls*
- *rubber or plastic tubing*

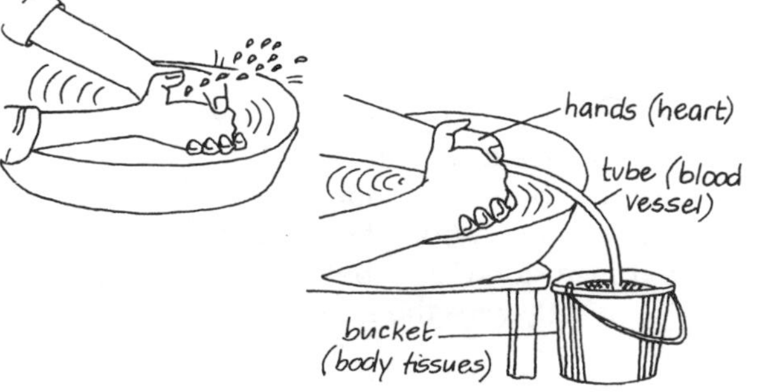

Open and close your hands as shown while they are in a bucket or bowl of water. Now hold a rubber tube as shown. Open and close the palms again. Opening and closing of the palms represents the relaxation and contraction of the heart muscles.

Heart model

You will need
- *cardboard box*
- *thin paper*
- *glue*

Make a model of the heart from a cardboard box as shown. Thin paper is used for the valves.

Discuss the importance of valves in the pump action of the heart.

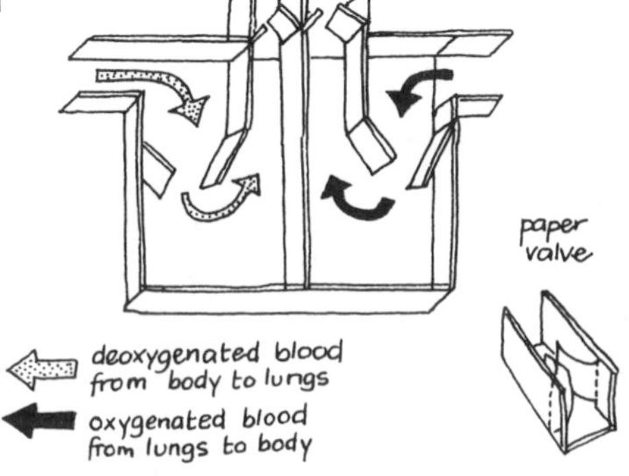

Blood vessels

Looking at blood vessels

The blood capillaries in the corner of the eye clearly show capillary size. Red meat has so many tiny capillaries they give it its colour.

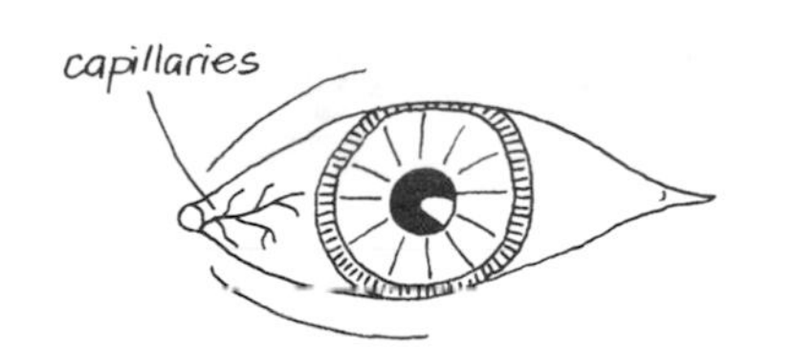

Blood vessels – a model

You will need

- *2 coloured ropes or thick string (one coloured blue, the other red)*

Untwist an end of each coloured rope until each end becomes a mass of tiny thin strings. If you twist the thin strings together they form a mass of fine capillaries.

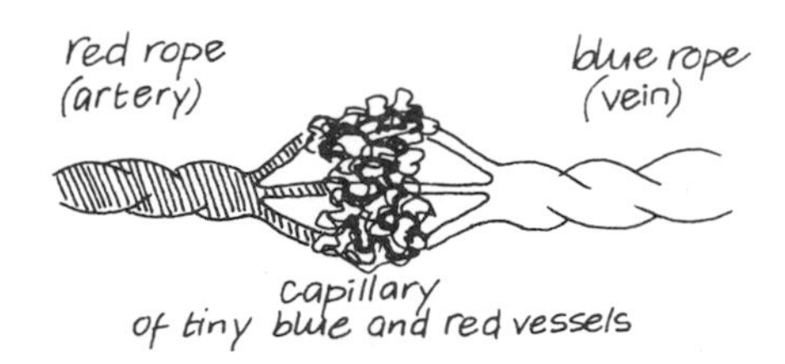

Circulation game

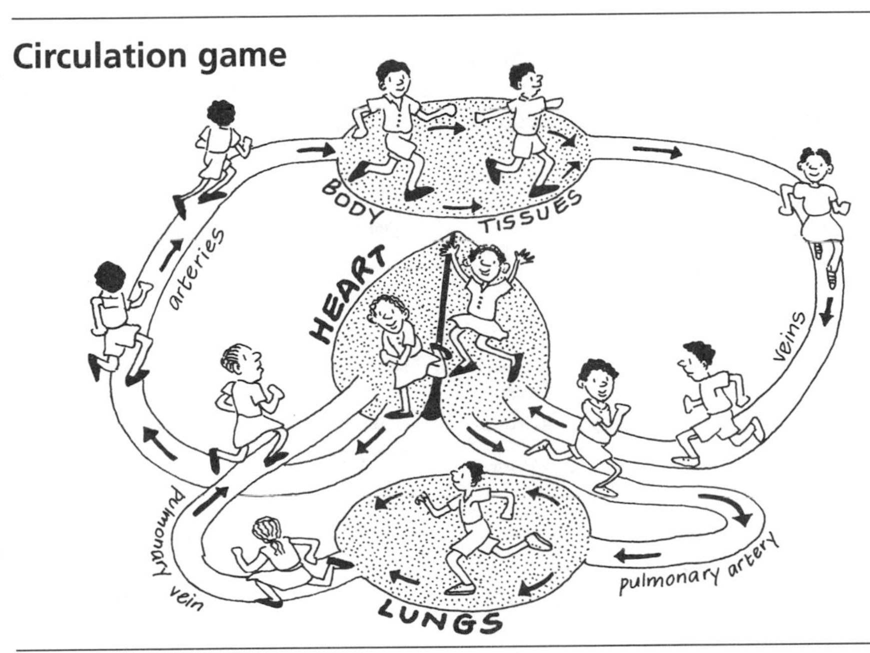

Peg or draw out a map of the circulation system on the classroom floor or school field. Students, representing the blood, walk around the system explaining what happens at each point.

Label each area or ask students to do so.

The human circulation system is called a double circulation system. Ask students why it is given this name and discuss the advantage of such a circulation system.

Ask students what happens if the heart beat rate increases?

Circulation overlay chart

Draw the base diagram on card or paper and all others on clear plastic. An A4 version could use clear plastic bags for the overlays. When all the drawings are placed over each other the full diagram is revealed.

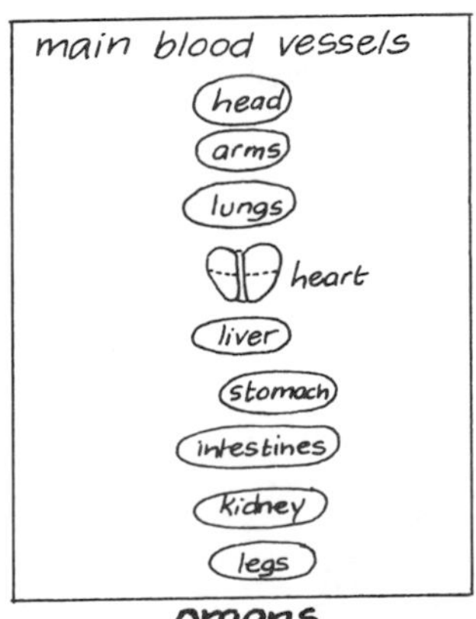

organs
(on base board)

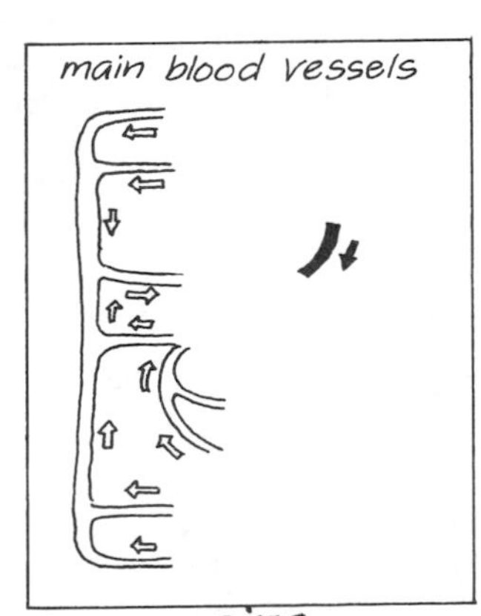

veins
(drawn on clear plastic)

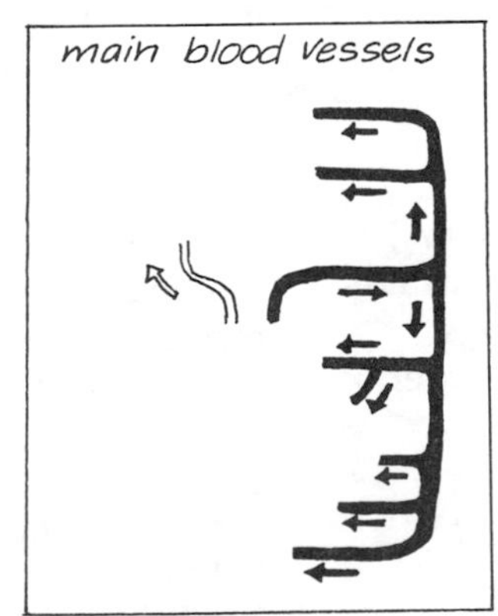

arteries
(drawn on clear plastic)

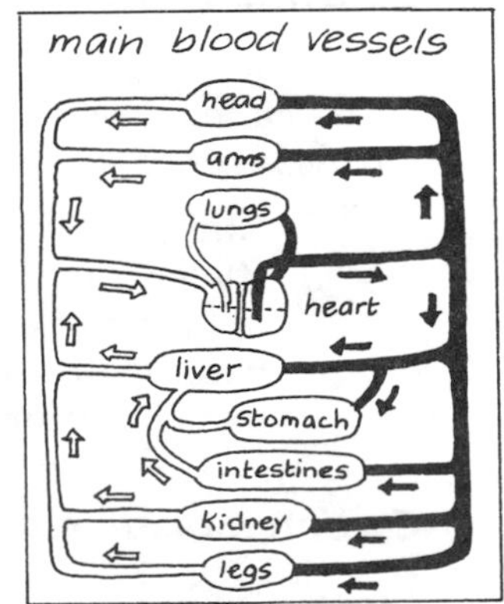

overlays together

Breathing

- **Breathing** is the process of drawing air into, and out of, the lungs
- Air is forced into the lungs because as the muscles of the **diaphragm** and the chest contract they enlarge the chest cavity. The resulting increase in volume reduces the **pressure** inside the lungs and so air rushes into them.
- Air breathed in contains more oxygen and less carbon dioxide than **expired** air because oxygen enters the blood from the lungs and carbon dioxide enters the lungs from the blood.
- **Gaseous exchange** occurs through the **membrane** of the lungs, which is folded into millions of tiny sacs, the **alveoli**.
- **Respiration** is the **metabolic process** whereby oxygen releases the **energy** in food with the production of carbon dioxide (see page 36).

Lung capacity

You will need
- *large plastic bag*
- *bucket*
- *large dish*

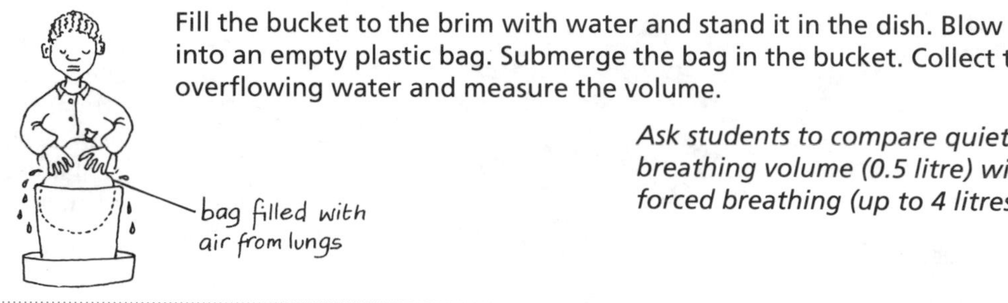

Fill the bucket to the brim with water and stand it in the dish. Blow into an empty plastic bag. Submerge the bag in the bucket. Collect the overflowing water and measure the volume.

Ask students to compare quiet breathing volume (0.5 litre) with forced breathing (up to 4 litres).

Diaphragm at work

You will need
- *a plastic bottle*
- *balloon*
- *plastic bag or rubber sheeting*

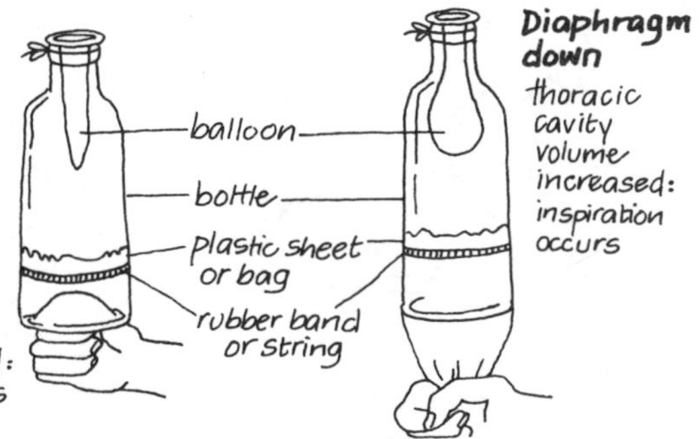

Cut the bottom off a plastic bottle. Attach a balloon over the bottle mouth so it hangs inside. Tie a piece of plastic bag over the cut base end.

Ask students to describe what they see using the terms pressure, volume, inspiration, expiration.

Investigating expired air

Moisture content

You will need
- *ballpoint casing or transparent straw*

Safety: **Do not allow students to share straws or ballpoint casings because of the danger of cross-infection.**

Breathe in through an empty ballpoint casing. Breathe out through the casing and notice the moisture droplets. The same effects will be seen by breathing out into a transparent plastic bag.

Ask students whether the droplets form on inspiration or expiration and where the moisture comes from.

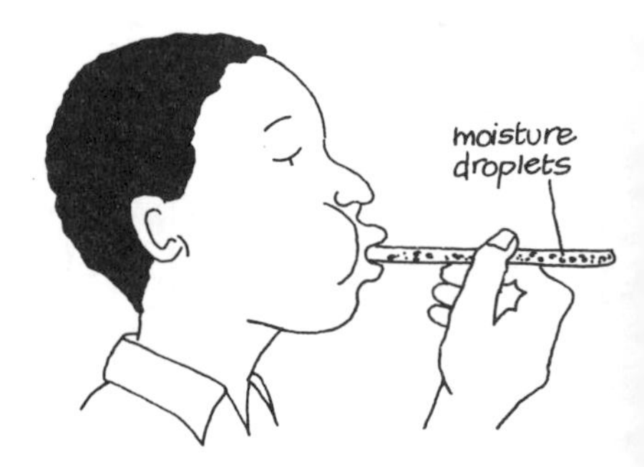

Carbon dioxide content

You will need
- *ballpoint casing or straw*
- *limewater*

Safety: **Do not allow students to share straws or ballpoint casings because of the danger of cross-infection.**

Use a straw or ballpoint casing to bubble air through the limewater. The limewater goes milky – the test indicating the presence of carbon dioxide.

Compare inspired and expired air by sucking in through the lime-water or blowing out through it.

Surface area increase

You will need
- *paper*
- *scissors*

Draw a circle on the ground to represent a lung. Find the circumference of the 'lungs', i.e. its surface area. Cut out many smaller circles to represent the air sacs. There are about 350 million air sacs in each human lung.

Ask students how many small circles can be fitted into the large circle.

What is the total circumference of all the small circles?

What effect does dividing the lung into air sacs have on the surface area of the lung?

Model of an alveolus

You will need
- *red and blue paint or waterproof markers*
- *inflated balloon*

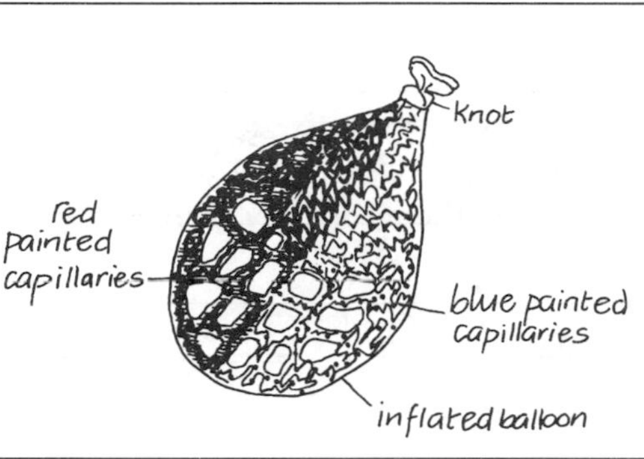

Use paint or waterproof markers to draw the blood vessels of the air sac or alveolus. The red vessels carry oxygenated blood, the blue vessels carry deoxygenated blood.

Gaseous exchange board game

You will need
- *large sheet of paper*
- *bottletops*
- *seeds*
- *stones*

blood from body (more CO_2 less O_2)
AIR IN
AIR OUT
blood to body (less CO_2 more O_2)
upturned top = deoxygenated red blood cell
top with stone in = oxygenated red blood cell
stones (oxygen)
small seeds (carbon dioxide)
capillary

Draw the capillary and alveolus as shown. Students arrange the stones (oxygen), seeds (carbon dioxide) and bottletops (red blood cells) on the drawing. The stones can be carried by overturned bottletops, but the seeds can only move in the plasma.

Gaseous exchange game

You will need
- *cards or paper marked with the letters R, P, O, CO_2*

The table represents the alveolus. Students wear either an 'R' or 'P' card and so act as red blood cells (R) or plasma (P). When going round the table the 'R' students pick up cards with 'O' (oxygen) on. The 'P' students put down the 'CO_2' (carbon dioxide) cards.

As an extension, link this with the circulation game – see page 33.

Respiration

- All living cells require energy to carry out essential processes, i.e. to live.
- Plants can trap the energy from the Sun by **photosynthesis**.
- Animals use the energy trapped in food. They do this by the processes of **respiration** or, in some cases, **fermentation**.
- All the chemical activities taking place in an organism are called the **metabolism** of the organism.
- **Waste products** of metabolism are poisonous and must be removed, e.g. we breathe out waste carbon dioxide and remove waste urea in our urine.

Cell respiration equations

Respiration cards

You will need
- *card*

Cut out cards to represent the substances involved in respiration. Label some cards with a '+' or an arrow. Mix up the cards and ask students to arrange them correctly as shown.

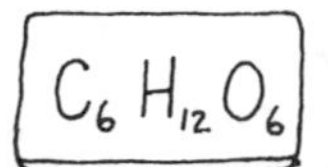

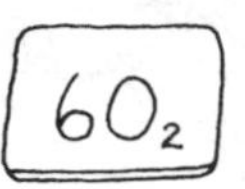

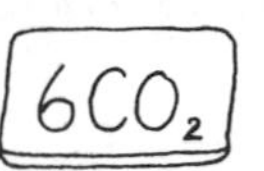

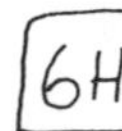

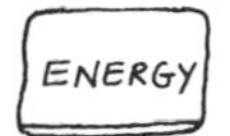

Respiration plates

You will need
- *a variety of seeds, bottletops, seeds, coins*
- *plates*
- *card*

Choose 3 different types of seed, coin or bottletop to represent carbon, hydrogen and oxygen. Arrange 4 plates or boxes on a table as shown. Ask students to place the correct number of seeds etc. on the plates. When the seeds are placed correctly the card carrying the 'E' for energy is added.

Discuss or demonstrate that the reverse equation is the process of photosynthesis.

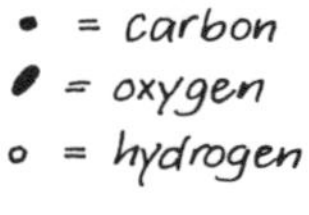

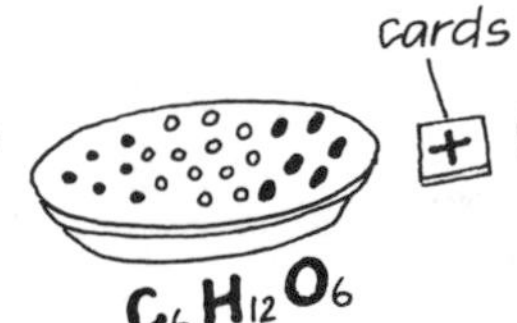

How much energy in a peanut?

You will need
- *peanut*
- *pin*
- *Plasticine or clay*
- *test tube*
- *flame*

Put 20 cm³ water into a test tube and record its temperature. Stick the peanut onto the pin and make a Plasticine or clay base. Place the peanut in a flame so that it starts to burn. Put the burning nut under the test tube. Record the temperature of the water after the nut has burnt away. The energy in the peanut can be calculated as shown:

amount of water (cm³) x rise in temp. (°C) x 4.2 = energy (Joules)

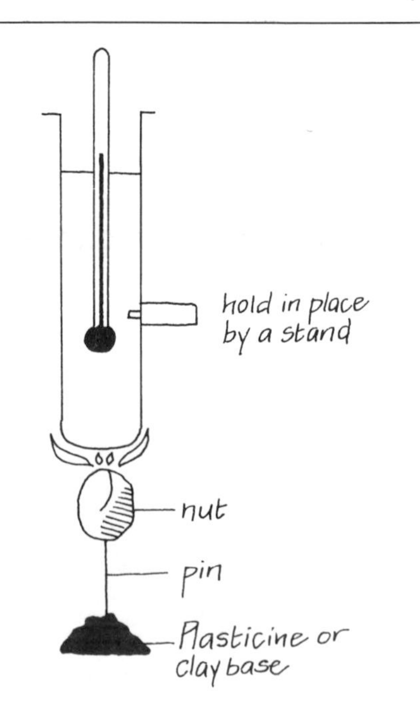

Respiration and heat

Measuring temperature change during germination

You will need
- *2 thermos flasks*
- *peas or beans*
- *2 thermometers*
- *cotton wool*

Place soaked germinating peas or beans in a thermos flask. Gently push a thermometer into the top and seal it with cotton wool. Set up a control test with boiled peas or beans. Record the rise in temperature of the germinating peas and note that the temperature does not rise in the dead (i.e. boiled) peas. The heat is a result of respiration taking place as the germinating peas grow.

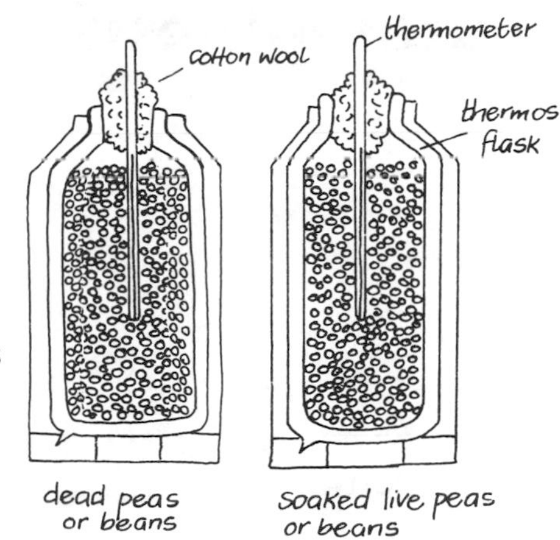

Hot grass!

You will need
- *newspaper*
- *freshly cut grass or leaves*
- *2 boxes*

Fill a box with freshly cut grass or leaves and pack them down tightly. Fill a similar box with the same amount of strips of newspaper. The leaves will produce heat but the control box of newspaper remains unchanged. The heat is produced by bacteria as they break down food (leaves, grass etc.) by respiration.

Respiration and carbon dioxide

You will need
- *net bag*
- *seeds*
- *limewater*
- *jar with lid or seal*

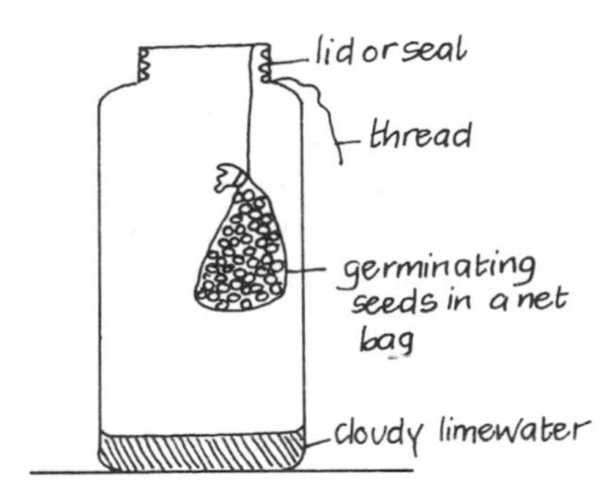

Hang a net bag filled with soaked and germinating seeds in a jar containing limewater. Seal the jar. Note that the limewater eventually becomes cloudy, indicating that carbon dioxide has been produced by the seeds.

Try the experiment using dry or boiled seeds.

Fermentation

Using yeast

You will need
- *yeast*
- *sugar*
- *limewater*
- *bottle and tubing as shown*

Place a little yeast in a solution of sugar and water. Test the gas produced by bubbling it through limewater. The limewater should go milky, indicating carbon dioxide has been produced.

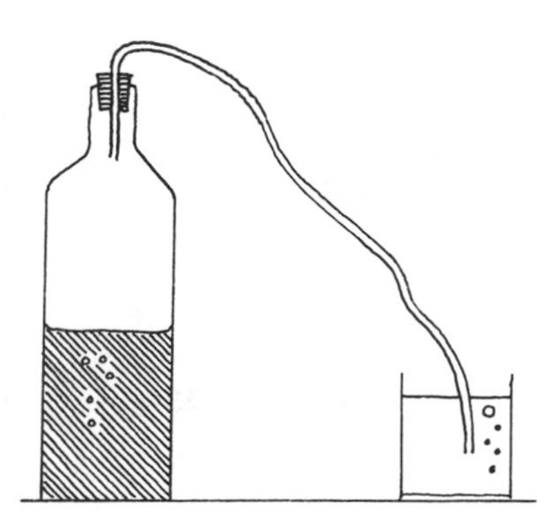

Using fruit

Cut up some fruits and place them in a jar or cup. Let the jar stand in a warm place for a day or so. The fruit ferments, acted on by wild yeasts found on the skins of the fruits.

38 Photosynthesis

- Plants make their food by **photosynthesis**.
- During photosynthesis carbon dioxide and water, in the presence of light and **chlorophyll**, are changed into glucose and oxygen.
- The glucose is changed to starch so the presence of starch indicates photosynthesis has occurred.
- In most plants a green **pigment** called chlorophyll is essential for photosynthesis to occur.
- Chlorophyll is found in the **chloroplasts** and these are more concentrated in particular cells, e.g. in the leaves.

Leaf structure

Leaf cells

You will need
- *strips of paper*
- *box (optional)*

Construct the different types of cells in the leaf from paper strips. Arrange the 'cells' in a leaf. This is easier if you put them into a surround or a box with sides. Pieces of grass can be used to represent chloroplasts.

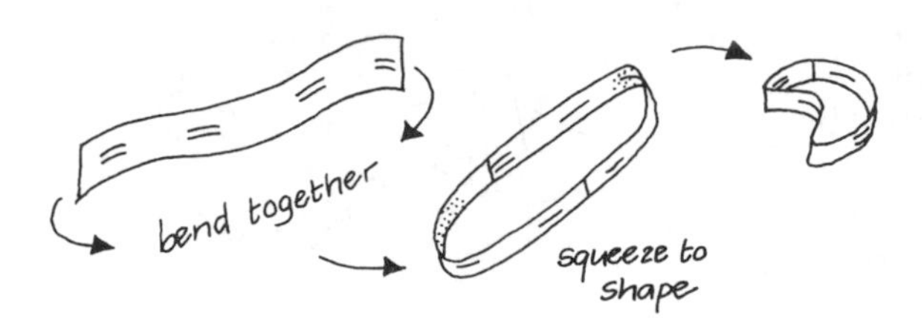

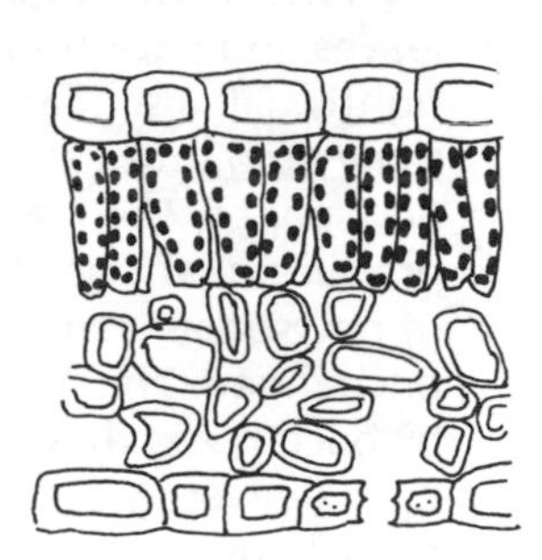

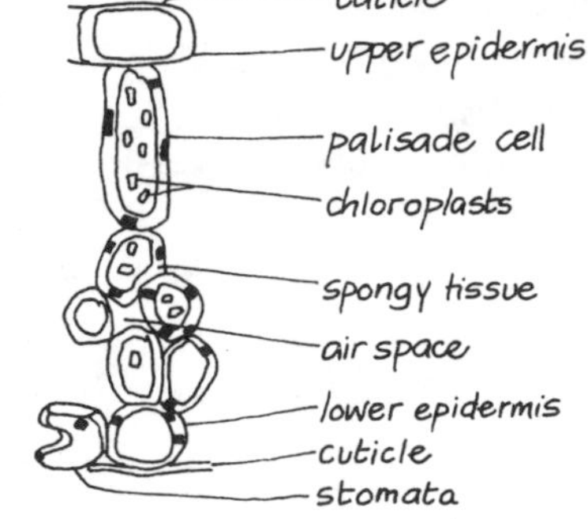

Making a model leaf

You will need
- *box*
- *coloured crayons or paints*

Paint or draw on the sides, top and bottom of a box. Remember that the inner surface of a printed box has a blank surface! It is easier to do the drawing with the box opened out flat and then reglue it into shape.

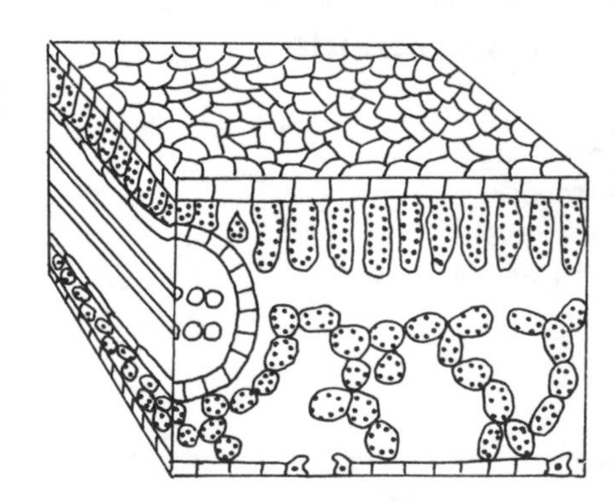

Exploring photosynthesis

The chemical equation for photosynthesis is

$6CO_2 + 6H_2O \xrightarrow{sunlight} C_6H_{12}O_6 + 6O_2$ in the presence of chlorophyll

Photosynthesis made easy

Draw and cut out cards of the shapes shown and then arrange them to show the chemical process of photosynthesis.

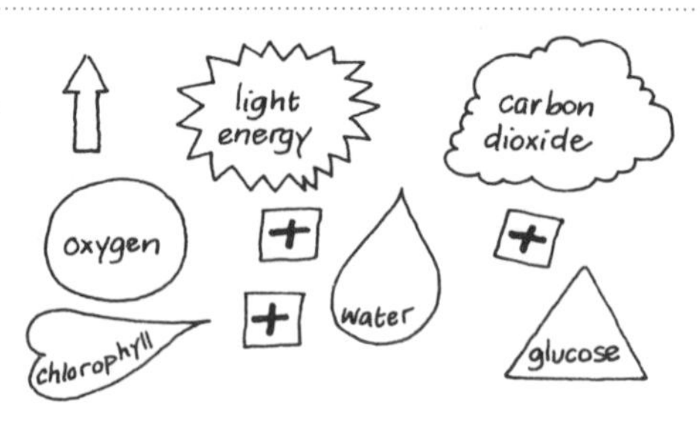

Photosynthesis equation game

You will need

- *appropriately-sized beans, coins, stones, bottletops etc.*

Arrange the items so they represent the stages of photosynthesis as shown in the diagram.

Ask students to adapt this to make it a game for 2 or more players.

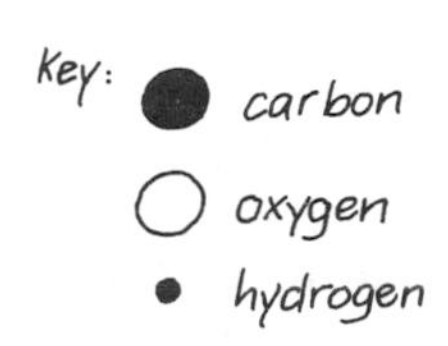

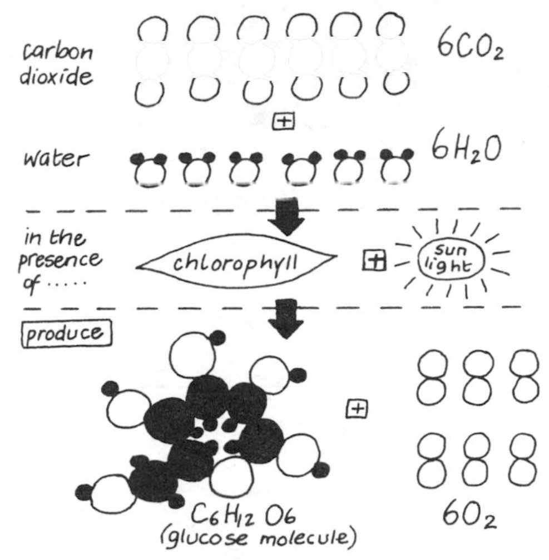

Light and photosynthesis

Using grass

Cover an area of grass with a large flat stone, cardboard, or black plastic so that no light reaches the plants. Notice the covered plants become yellow and eventually die. A test for starch would show the leaves have not produced starch, i.e. the plants have not been photosynthesising.

Using pot plants

You will need

- *2 pot plants*
- *dark cupboard*
- *alcohol*
- *source of heat*
- *iodine solution*

Safety: **Alcohol is a fire hazard, so use a water bath to heat it.**

Take 2 pot plants. Place one in sunlight and the other in a dark cupboard for 2–3 days. Pick a leaf from each plant and remove the green colour by heating the leaves for about 5 minutes in alcohol.

Test each leaf for starch (see Food Tests page 27.)

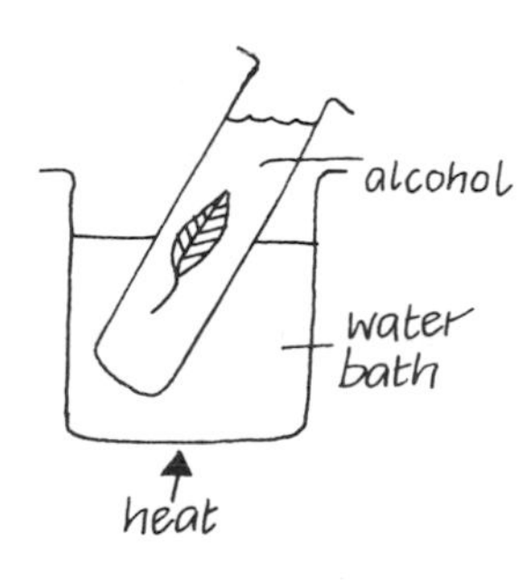

Chlorophyll and photosynthesis

You will need

- *variegated leaf*
- *alcohol*
- *water bath*
- *source of heat*
- *iodine solution*

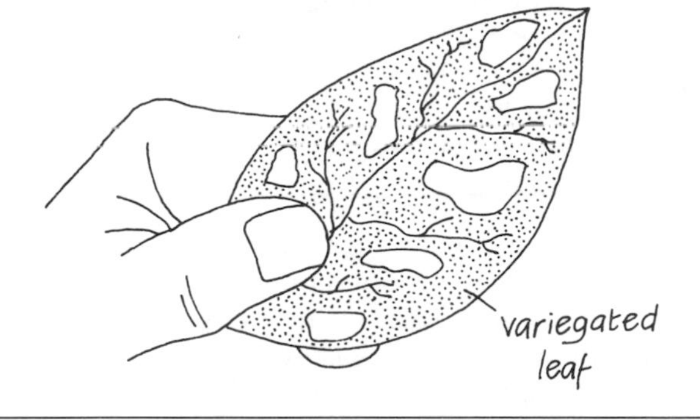

Find a leaf which is not all green, e.g. ice plant (Breynia) or nasturtium (Croton). Draw the leaf, carefully identifying the green areas where chlorophyll is present. Test the leaf for starch. (Boil the leaf in alcohol first). Notice that the areas which turn blue-black during the test are the areas of the leaf which were green.

Carbon dioxide and photosynthesis

You will need

- *a plant which is not growing in direct sunlight*
- *clear plastic bag*
- *rubber band or wire*
- *pellets or liquid sodium hydroxide*
- *alcohol*
- *water bath*
- *source of heat*
- *iodine solution*

Sodium hydroxide absorbs carbon dioxide.

Place a clear plastic bag over one leaf of a plant as shown and leave it for a day. Test the leaf in the bag for starch and also test another on the plant. (Boil leaves in alcohol before testing for starch.) The leaf which has been in the bag will not have starch in it, i.e. no photosynthesis has taken place.

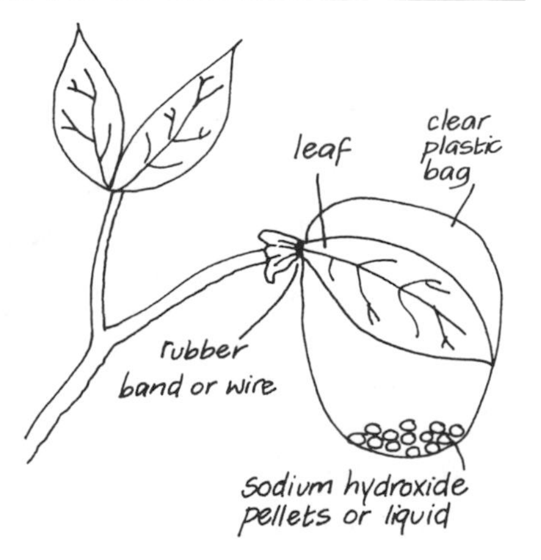

40 Plant transport and transpiration

- Plants have **hydrostatic skeletons**, i.e. it is the **pressure** of water in the **cells** which keeps the plant rigid.
- Water enters through the roots and moves through the **xylem vessels** up to the leaves where it is lost through small holes, the **stomata**.
- Food, made in the leaves, is transported in the **phloem** to other parts of the plant.
- **Essential nutrients** are absorbed through the roots by **active transport** and then carried to other parts of the plant in the xylem.

Leaf structures

You will need
- *a variety of different types of leaves*
- *sheet of clean white paper*
- *paint or ink*

Examine the different types of leaves and notice in particular the patterns of the leaf veins. Make leaf prints by lightly painting or inking the surface of a leaf, placing a piece of paper on top and then pressing down lightly.

Students could make a leaf-print collection.

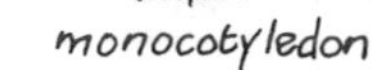

Looking at stomata

The distribution of stomata on leaves

You will need
- *glass container full of hot water (not as hot as boiling)*
- *a variety of leaves*

Place a leaf in hot water. Bubbles come from the stomata as the air in the leaf expands. Look at different types of leaves and notice the distribution of the bubbles, i.e. the stomata.

Students will see that in monocotyledons the bubbles emerge from both sides of the leaf. Investigate whether this is true for other types of leaf too.

Examining stomata in detail

You will need
- *leaves*
- *microscope*
- *microscope slides*
- *nail varnish (optional)*

Snap a leaf in half and carefully peel off a portion of the lower epidermis. Place the epidermis on a slide and examine it under a microscope. The curved guard cells will be clearly visible.

Alternatively, lightly coat the underside of a leaf with clear nail varnish. After it has dried peel it off and the impression of the cells will remain in the nail varnish. Note that this will work better with some leaves than others.

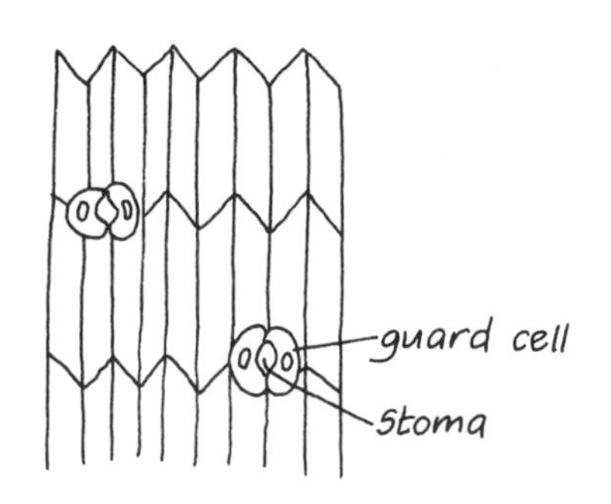

Transpiration

You will need
- *potted plant*
- *2 small plastic bags*
- *string*
- *grease or Vaseline*

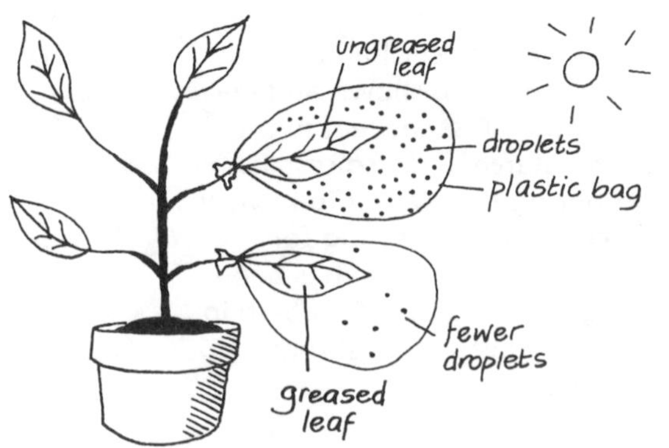

Water loss through transpiration is clearly visible in transplanted seedlings on a hot day.

You can demonstrate that the water is lost through the stomata by comparing greased and ungreased leaves. Grease both sides of one leaf and place a separate plastic bag around each leaf. Compare the amount of water in the plastic bags later.

Looking at roots

Dig up roots from a variety of different types of plants and compare them.

Discuss with students whether their specimens have fibrous or tap roots.

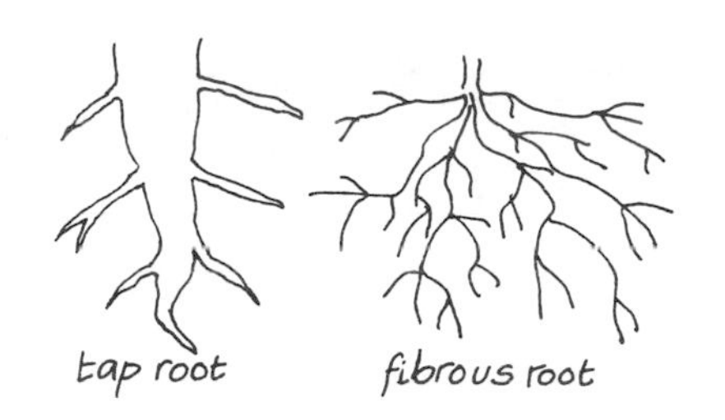

Looking at root hairs

You will need
- *peas or beans as seeds*
- *damp cloth or paper*

Germinate some seeds on a damp cloth and examine the root hairs. Root hairs increase the surface area for absorption. Most of the absorption of water occurs in the fine root hairs.

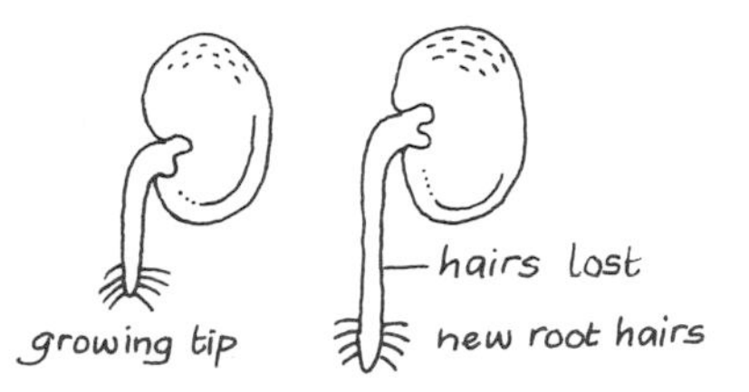

Model: active transport

You will need
- *matchboxes*
- *dried peas*
- *bottletops*

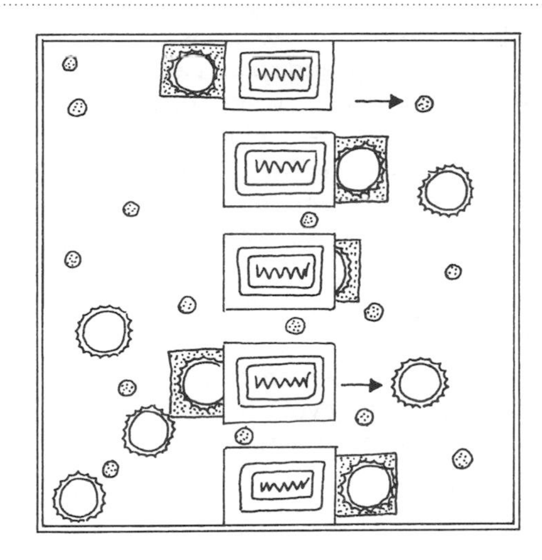

Make a model as shown in the diagram. Note the matchboxes are lined up and the drawers can be opened either side. The matchboxes and spaces between them represent a selectively permeable membrane. The spaces allow small objects through, but not larger ones. The peas represent water molecules which move freely. The bottletops represent the larger glucose molecules which need to be placed in the matchbox drawers and actively pushed through to the other side.

Tracking water movement

You will need
- *a selection of different plant stems*
- *coloured (i.e. not black) ink or dye*
- *water*
- *sharp knife*

Place a variety of different types of plants in coloured ink or dye and leave them for a few hours. Slice off sections of the stem with a sharp knife and examine them under a handlens. The colour is located in the xylem vessels which shows water is transported in the xylem.

Some very young plants, such as Balsam, are so transparent that you can see the colour move up the stem.

Ask students to note the distribution of colour in different plants and compare roots and stems.

Vascular tissue game

Chalk 2 circles on the floor or table. Cut out 20 discs from card or paper. Colour 10 to represent xylem vessels and 10 to represent phloem tubes. Arrange the discs to show the arrangement of vascular tissue in a root and a dicotyledon stem.

Ask students if they can adapt the game.

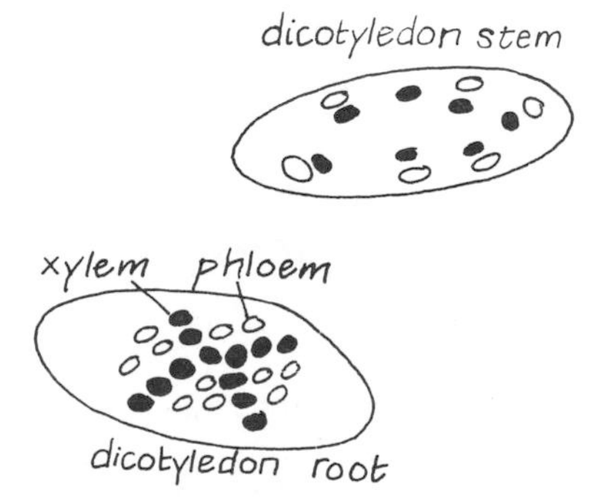

Support and movement

- Our **skeletons**, like those of many animals, are made of **bone**.
- **Muscles** are attached to the skeleton.
- **Joints** allow movement between bones, each type allowing movement in a particular direction.
- Joints, and limbs, are moved by muscles.
- Muscles can only pull, not push and so muscles always occur in **antagonistic pairs**.
- Muscles not only give movement, they also support parts of the skeleton.

Joints

Joints prevent the ends of articulating bones wearing away. The end of the bone is covered by a layer of cartilage, which is slightly springy and so acts as a shock-absorber. Between the two layers of cartilage is a lubricating fluid, the synovial fluid. Collect animal bones from butchers to demonstrate particular joints.

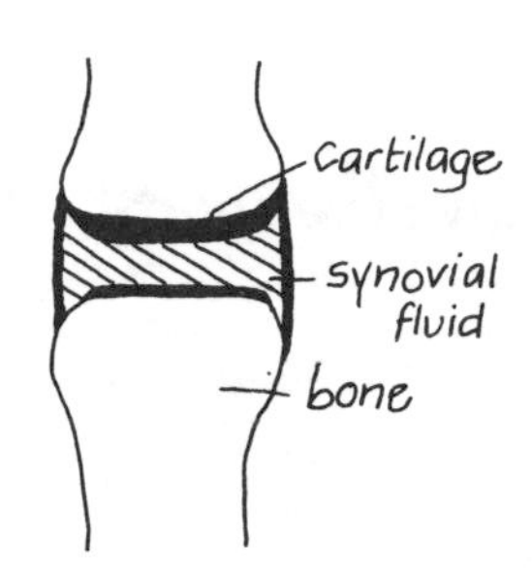

Ball and socket

You will need
- *light bulb*
- *coconut shell*
- *stick*

The hip joint, which allows the thigh to move, is a ball and socket joint. You can demonstrate such a joint by cupping your hands or making one as shown.

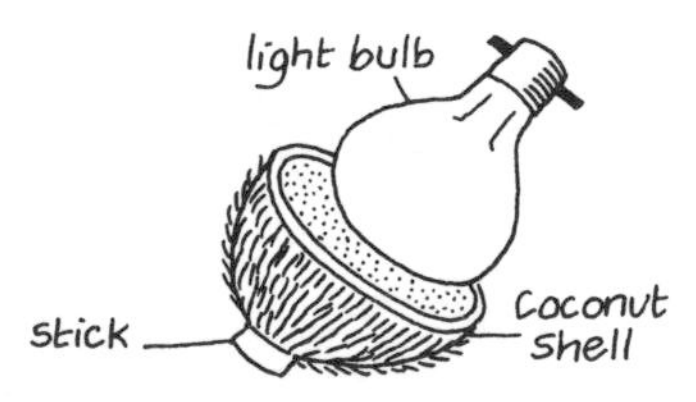

Hinge joint

You will need
- *stick*
- *round piece of wood or a can*
- *plastic bottle or a can*

The elbow and knee are both hinge joints and allow movement in only one direction – like a hinge. You can make a model of a hinge joint as shown.

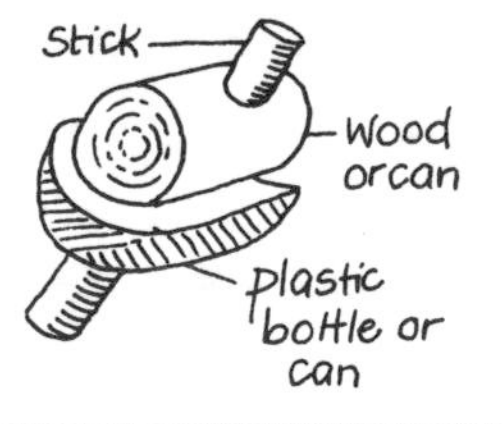

Sliding joints

You will need
- *string*
- *cans or cotton reels*
- *sponge or card*

The joints between vertebrae allow movement of the spine. Make a model of the spine as shown.

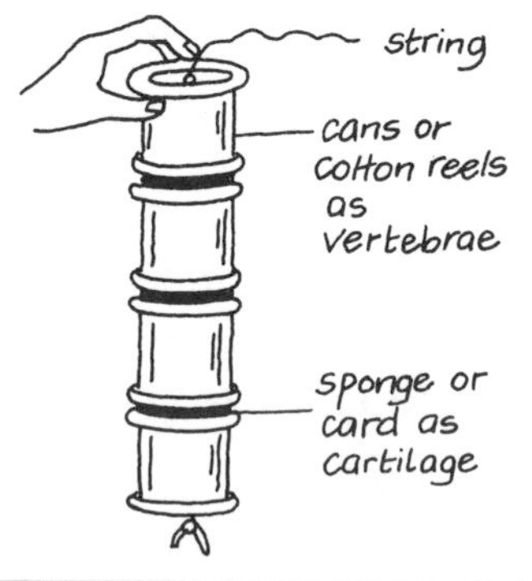

Minerals in bones

You will need
- *vinegar or acid*
- *a bone (chicken works well)*

The disease rickets, a symptom of which is bowed legs, is a result of soft bones. If our diet lacks vitamin D our bodies cannot absorb calcium phosphate, an essential component of strong, rigid bones.

Bones left in vinegar for 2–3 weeks become soft and pliable because the acid removes the minerals that make bones strong and rigid.

The forearm as a lever

You will need
- *wood or card*
- *string*
- *2 strong straight sticks*

Make a model of the forearm as shown and notice that the arm can only be bent by shortening one 'muscle' at a time.

Ask students to make one model using rubber bands instead of string.

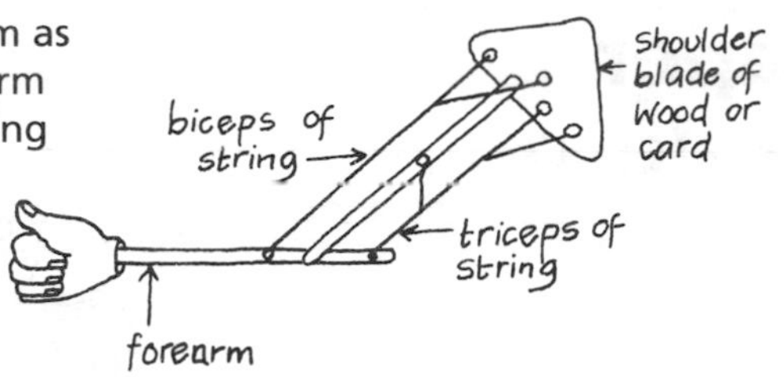

The effect of load on muscle

You will need
- *2 tins filled with sand*
- *ruler or scale*
- *rubber strip*
- *2 strong sticks*
- *weights*

Make a model arm as shown. Use a light weight to begin with and then increase the load.

Discuss what happens to the muscle as you increase the load (weights) on the lever (arm) and the effect of the position of the weight on the 'arm'.

Students should move their arms to correspond to the model. Discuss with them where they usually carry loads on their arms and why.

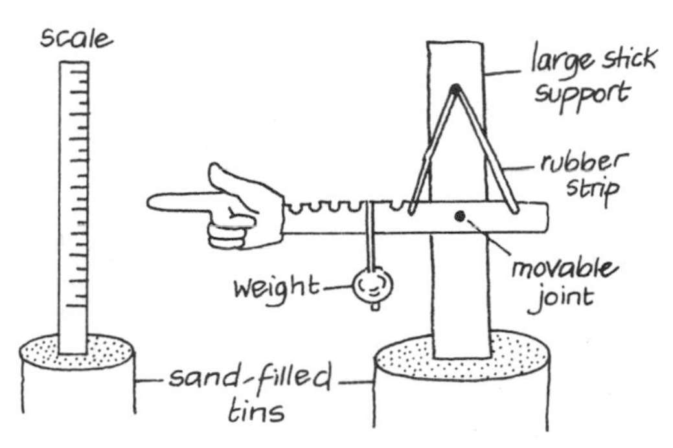

Tendons

You will need
- *chicken's foot*
- *tweezers*

Muscles are attached to bones by tendons. Some tendons may be quite long. Demonstrate how tendons pull on a joint by looking at a chicken's foot.

Expose the white, string-like tendons around the cut of the chicken's foot. Pull each tendon, one at a time and discuss the resulting action of the toes.

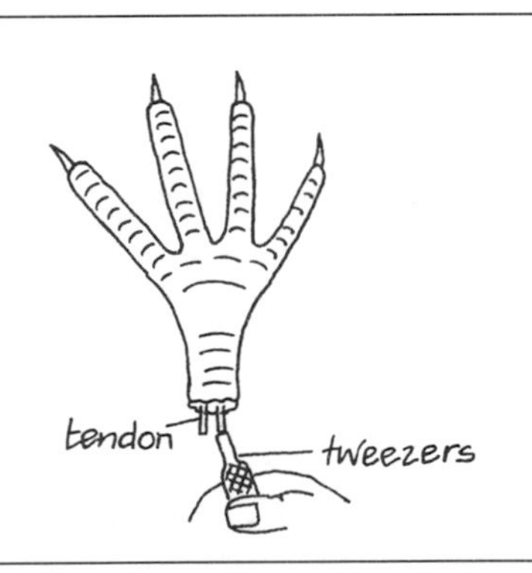

Support of the spinal column

The diagrams below show the position of the spinal column in relation to the legs.

Ask students to load the 'backbone' by adding weights to it and discuss the effect on the joints.

Discuss the role of muscles in maintaining the posture of each animal.

Muscles work in pairs

You will need
- *rod*
- *rope or thick string*
- *small tin or chalk mark on floor*

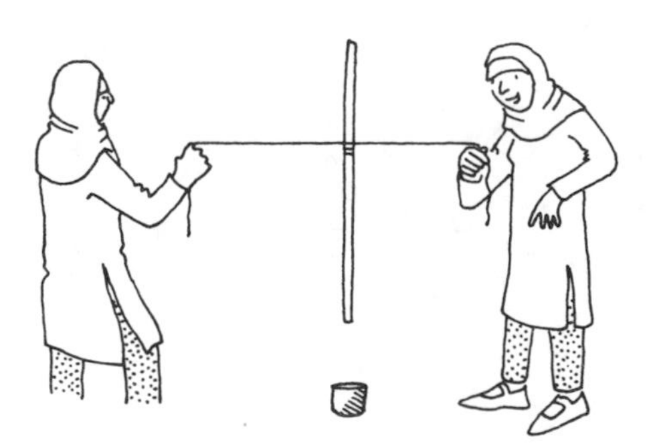

Tie the string to the rod as shown and ask pairs of students to manoeuvre the stick into the tin, or onto a chalk mark on the floor. The rope can only pull the rod, not push it. Muscles can only pull too.

Use a pencil for a desk-sized model.

Paper skeleton

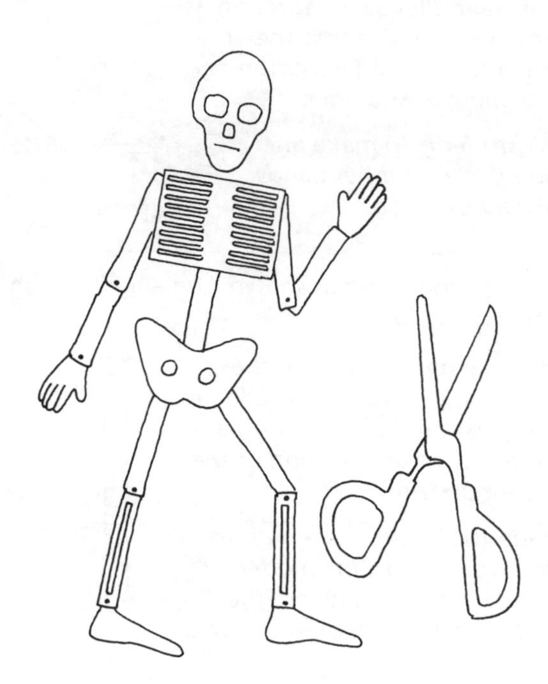

To make the paper skeleton shown here you will need 8 pages of A4 paper or pages from a large writing book used by students. Fold and cut out the shapes as illustrated for each part of the body. The final result should look like the one shown.

Skull

Cut around the dotted line after drawing. The teeth and mouth can be cut without removing any paper.

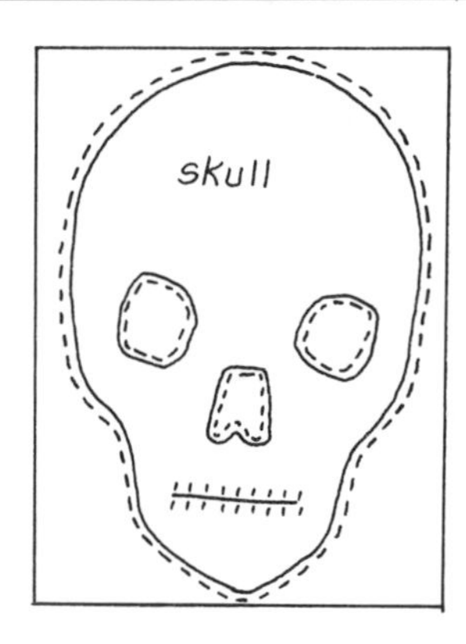

Pelvis and shoulder blades

Draw half of the pelvis and cut out the basic shape when the paper is folded. Cut out shoulder blades in the same way using an extra piece of paper.

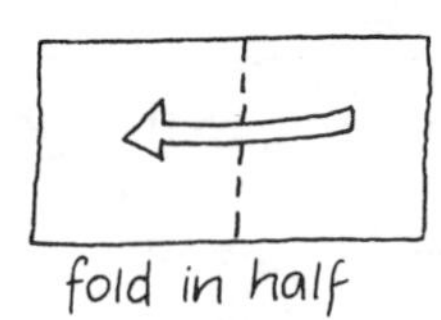

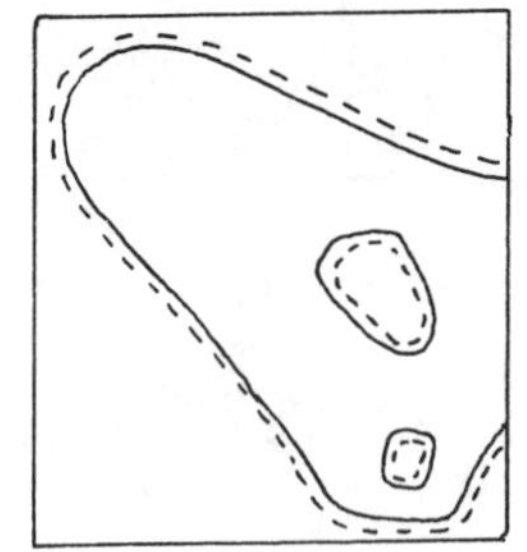

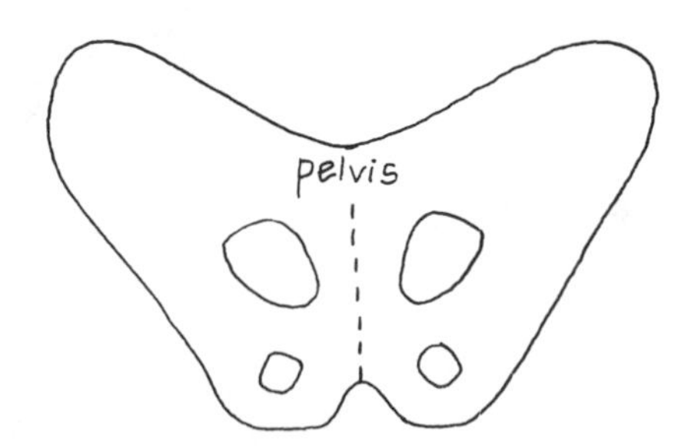

Limbs

The lower limbs are cut out from one piece of paper. The upper limbs all fit onto another piece.

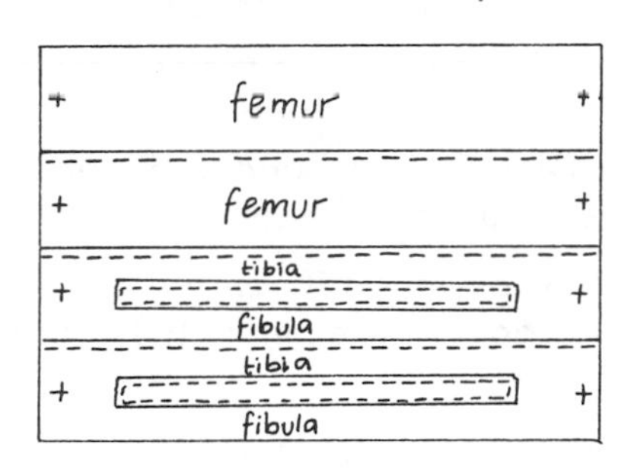

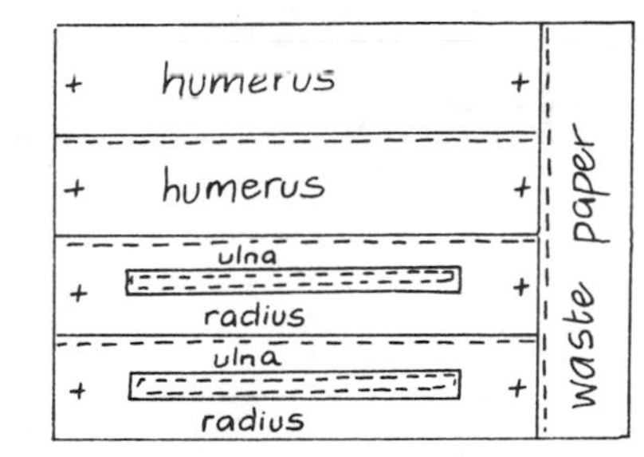

Rib cage

Fold the paper twice and then cut along alternate lines. Use a ruler to measure accurately if you want to have the exact number of ribs. You can cut the ribs out of the paper lengthwise instead.

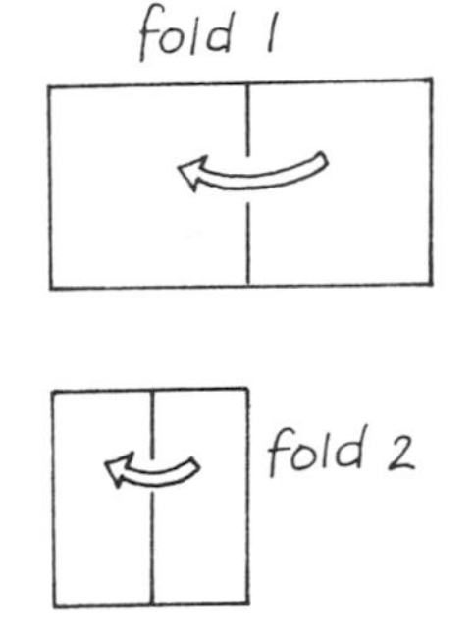

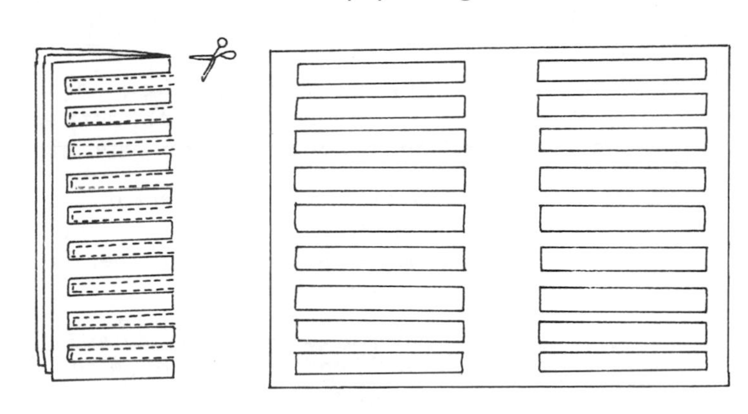

Backbone

Cut out 2 strips for the backbone to give extra strength. Stick one piece to each side of the skeleton.

Hands and feet

Fold the paper in half and ask students to draw around a hand. Use another piece of paper for the feet.

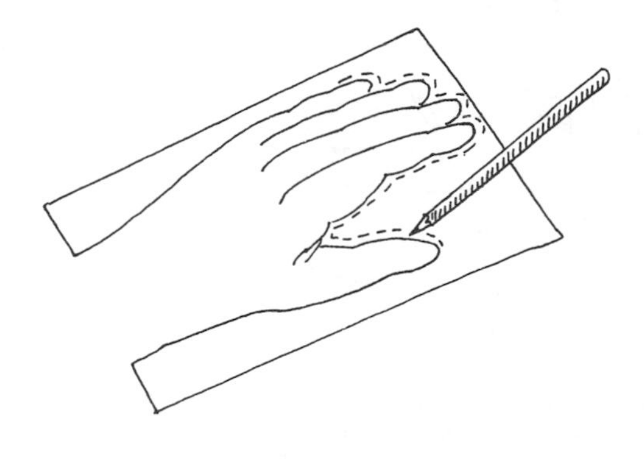

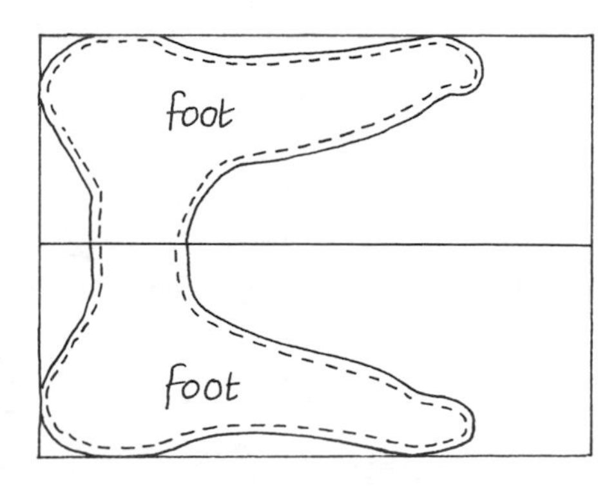

More ideas

- The bones of the feet and hands may be drawn onto the paper outlines.
- The paper limb bones can be shaped to add greater realism.
- Pin or staple the skeleton together or mount it on a hanging mat (see page 19).

46 Senses and responses – plants

- Plants **respond** to certain **stimuli** either by nastic movements or tropisms.
- Parts of some plants move in response to light or touch, e.g. petals close at night. Such movements are called **nastic movements**.
- Shoots and roots respond to stimuli by growing. These movements are called **tropisms**.
- The most obvious difference between tropisms and nastic movements is that tropisms take longer to happen.

Geotropism – movement with gravity

Shoots

Lean a pot plant at an angle. Leave it for a week. Notice that after this time the leaves turn upwards.

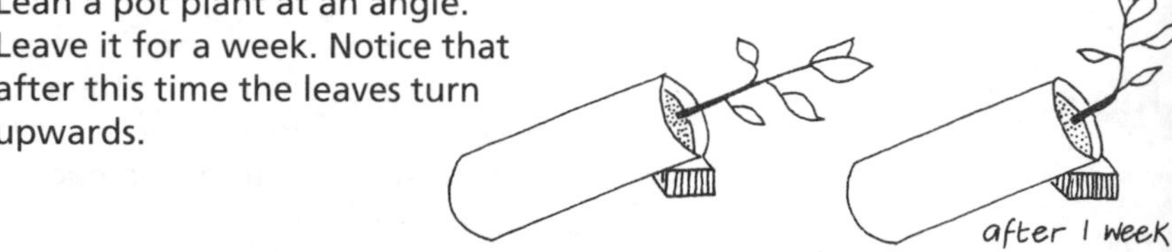

Roots

You will need
- *damp newspaper*
- *beans*
- *glass jar*

Put several seeds between the jar and the newspaper as shown. The seeds should be placed at different angles. The root will always grow downwards, towards gravity, and the shoot upwards.

Ask students if they think it matters which way up seeds are planted and why before they do the experiment.

After the experiment ask if the experiment supports the students' original views.

Phototropism

Farmers and gardeners see leaves turning to the Sun after disturbance or transplanting. Place a house plant next to a window letting in sunlight. Leave it for a few days. Now rotate the pot and note the position of the leaves. Examine the plant over the next few days. The leaves turn towards the light as the plant grows.

Find the light maze

You will need
- *cardboard box*
- *seedlings in small pots*

Make a light maze box as shown. Lift the lid daily to watch progress.

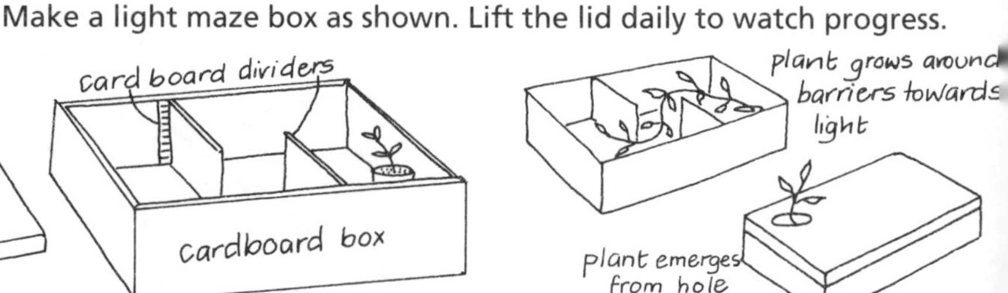

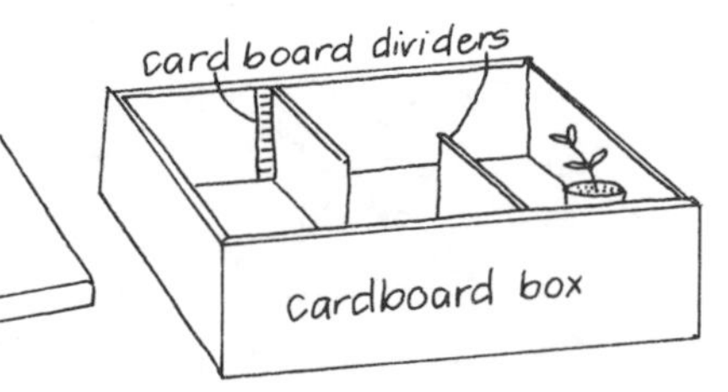

Sensitive stem tips

You will need
- *maize seedlings*
- *tin foil, sweet or cigarette paper*
- *cardboard box*

Cover the stem tips of the seedlings in one pot. To make the cap, roll the foil around the end of a matchstick. The foil must prevent sunlight reaching the tip. Compare the growth of the normal seedlings with those covered by foil caps. Put both pots of seedlings into the box. The seedlings without caps grow towards the hole, i.e. towards the light. The seedlings covered by caps grow straight up. This shows that only the growing tip is phototropic.

Hydrotropism – movement towards water

Porous pot

You will need
- *a large dish*
- *a porous pot*
- *soil*
- *water*
- *seedlings*

Fill the porous pot with water as shown. The seedlings' roots will grow towards the porous pot.

Try this with an empty porous pot.

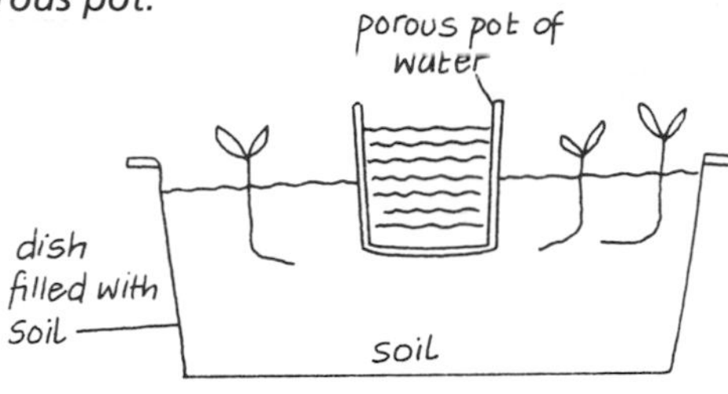

Searching for water

You will need
- *an empty box or tin*
- *fine mesh, e.g. mosquito net*
- *beans or peas*
- *soil*
- *saucer*

Cut a hole in the bottom of a tin or box. Fit a piece of wire netting (mosquito net) over the hole. Cover the netting with 2–3 cm of moist soil and plant one or two beans. Place the tin or box over an empty dish or saucer. The roots grow down at first (responding to gravity) but they turn to the side, searching for water. Now repeat the experiment with water in the saucer.

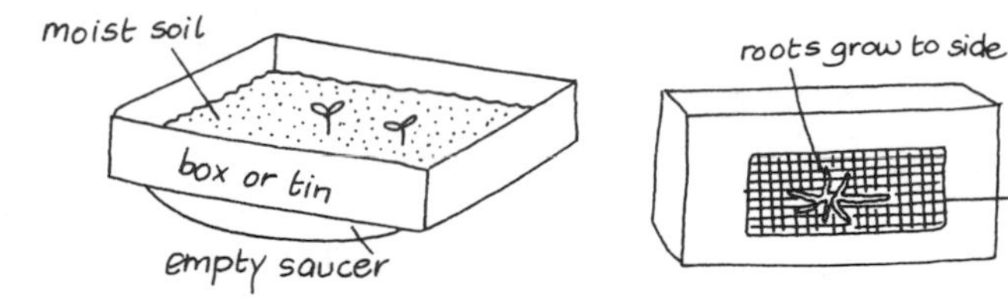

Temperature affects growth

You will need
- *beans*
- *2 glass containers*
- *damp newspaper*
- *cold place (fridge)*

Grow one glass of beans at room temperature. Keep the other glass as cool as possible. The cool beans will grow more slowly. In a fridge they may not grow at all.

Simple hothouses

You will need
- *plastic bags*
- *wire or stick supports*
- *plastic bottles*

Hothouses are warmer than the outside air and so crops, such as lettuce or tomatoes will grow faster.

Use plastic bags supported by sticks or wire to form a hothouse over any container.

Cut a door in a plastic bottle and plant seeds inside the mini-hothouse.

Nastic movements

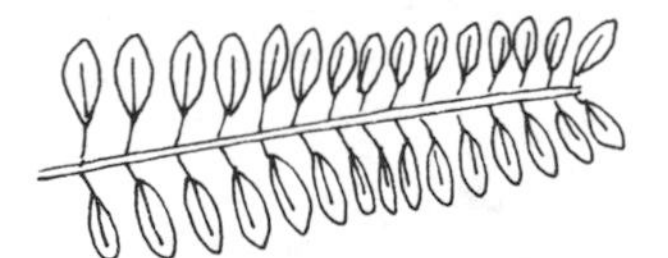

Look out for the following plant movements in your environment.

- Petals of flowers closing when it gets dark.
- Petals of flowers opening when it gets dark, e.g. Moonflower. (These flowers are often pollinated by night-flying insects.)
- Plants like *Mimosa pudica* (Sensitive Plant) which close up their leaves if touched lightly. If Mimosa is touched heavily the whole plant collapses.
- Plants which fold their compound leaves in a definite way, e.g. Sensitive Plant and Oxalis.

Senses and responses – animals

- Sense organs detect **stimuli**, e.g. the ear **detects** sound.
- Nerves carry **impulses** from **receptors** to the brain or spinal cord.
- A **reflex action** is an involuntary, very fast **response** to a stimulus.
- We often use several senses together, e.g. smell with taste. The sense of sight affects our balance.

Taste map

You will need
- *prepared taste solutions*
- *glass rods or matchsticks*

Prepare the 4 taste solutions as follows.
- Bitter – lemon peel, chloroquin (1 malaria tablet dissolved in a cup of water, but test before use), strong cold tea
- Sour – vinegar or lime juice
- Salt – salt dissolved in water
- Sweet – sugar dissolved in water

Use the solutions to map out the areas of taste on the tongue. Glass rods or matchsticks dipped into each solution allow a single drop to be placed accurately. The areas students should find are shown opposite.

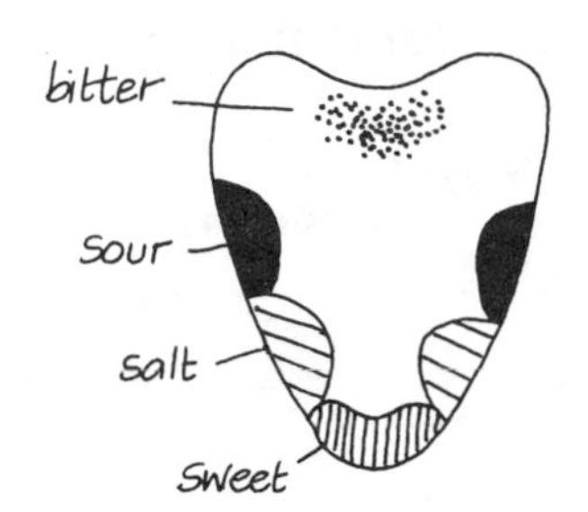

Ensure students are made aware of health hazards from using unclean or shared solutions. Rods or matchsticks should never be shared.

Nerve model

Use sticks, straws or grasses as substitutes for pencils.

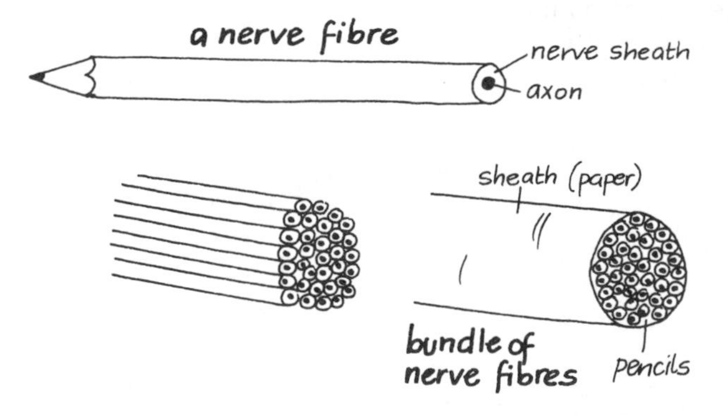

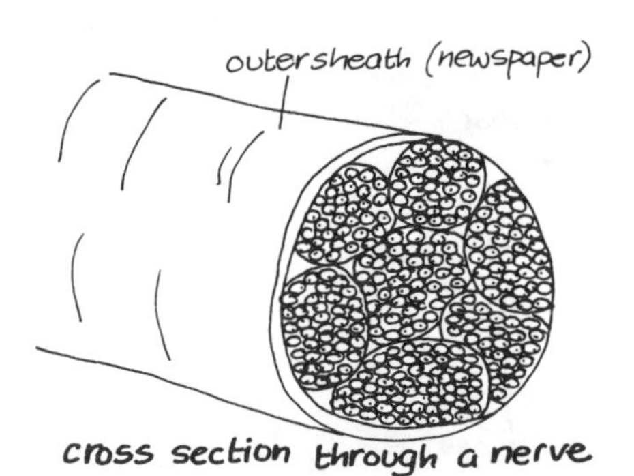

Reflex actions

Light and the eye

Cover one eye and look into a bright light. After the uncovered eye has adjusted to the light uncover the other eye. Quickly compare the sizes of the pupils in each eye. The size of the pupil is adjusted by a reflex action. In bright light the circular muscles of the iris diaphragm contract and the pupil becomes smaller.

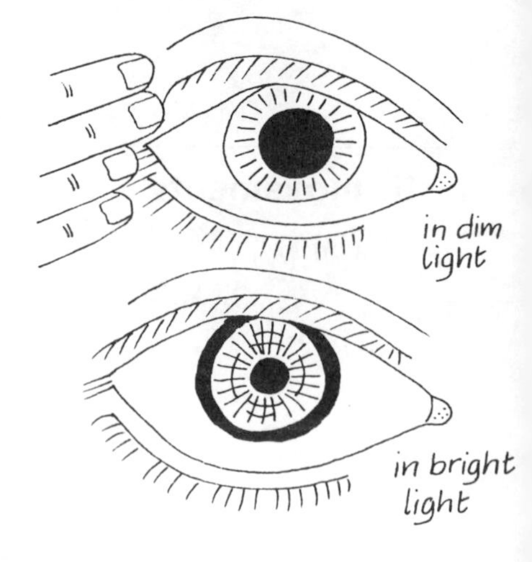

Blinking reflex

You will need
- *sheet of clear plastic*
- *small paper ball*

One student holds a clear piece of plastic to protect his or her eyes. The plastic from a large plastic bottle is suitable. Another student throws a crumpled ball of paper at the plastic. The first student blinks. Blinking is a reflex reaction.

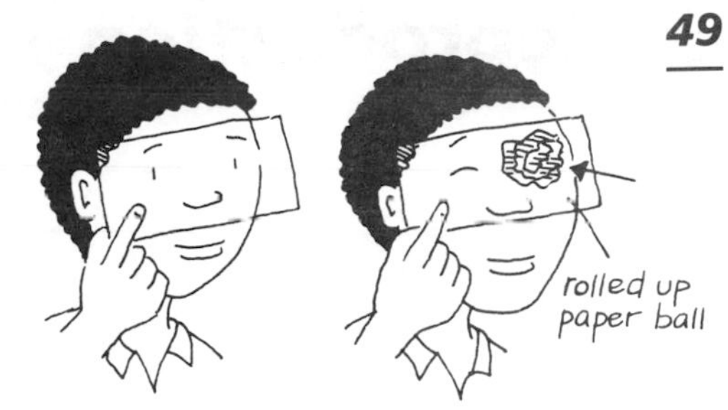

Knee jerk

Cross one leg over the other. Tap just below the knee cap as shown. The tapped leg kicks up in an involuntary reflex response.

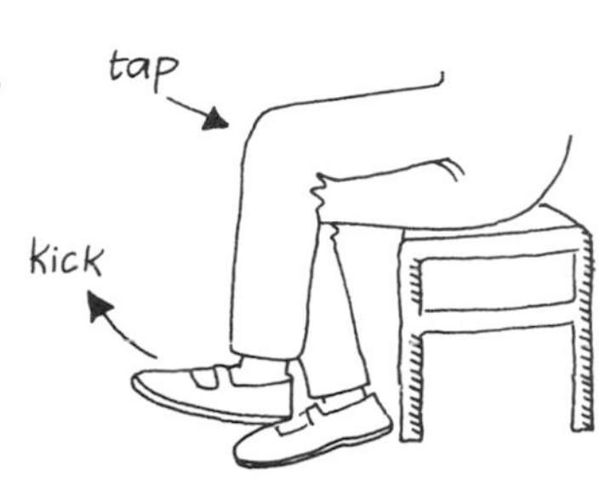

Linked senses

Sound and direction

One student is blindfolded and stands in a circle made by the others. One at a time each person in the circle makes a small noise. At each noise the blindfolded student points to the direction of the sound. Investigate students' hearing.

How accurately can students detect the direction of the sound?

Alter the distance between the sound and the detector.

Cover one ear (with cotton wool or a cloth).

Sight and balance

Try balancing on one leg with both eyes closed. Now try with the eyes open. It is easier to balance with the eyes open – sight is an aid to balance.

Ask students to spin round and discuss whether it is easier to regain balance when the eyes are open.

Tasteless food

You will need
- *a crisp apple*
- *an onion*

Cut an onion and an apple into small pieces. Blindfold the person to be tested and make them hold their nose. They will find the apple and onion taste the same! If students can detect the difference, try giving them the smell of onion as they eat the apple! Smell is very important in identifying foods. A cold or a blocked nose makes it difficult to taste properly.

Reproduction

- **Sexual reproduction** involves the joining of 2 single cells. One cell comes from the male, the other from a female.
- In humans the male cell is the **sperm**, the female cell the **ovum** (egg).
- During **fertilisation** the **chromosomes** from the male and female combine (see page 60) to form a **zygote**.
- **Flowers** are the sexual organs of plants and many flowers contain both the male part (**anther** producing **pollen**) and the female part (**ovary**).

Models of sperm and eggs

You will need
- *large football*
- *small bean with cotton tail*

The football represents the human egg, the bean a human sperm.

Ask students how much bigger the egg is than the sperm.

Sperm model

You will need
- *clay*
- *plastic bag*
- *adhesive tape*
- *fine thread*
- *string*

This model can be adapted.

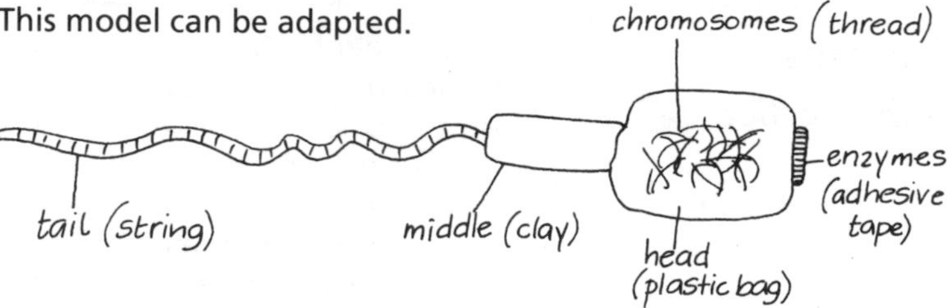

The egg's journey

Make a Plasticine or clay model of the female reproductive organs and discuss the path of the egg from the ovary.

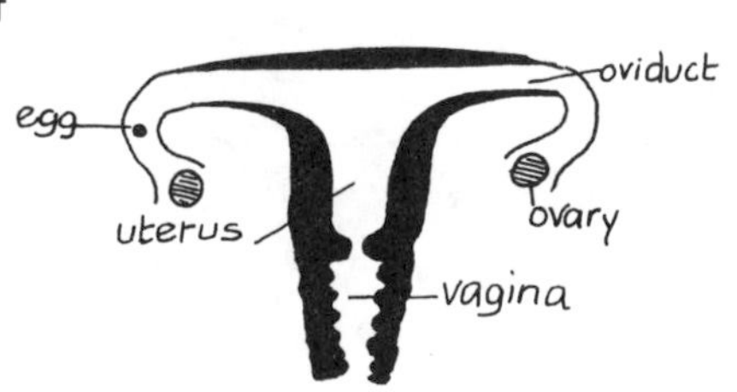

Fertilisation

You will need
- *2 small lids, e.g. bottletops*
- *fine thread*
- *string*
- *large plate*

Make the sperm and egg cell as shown. Note that the lids represent the nuclei of the female and male cells. The plate represents the egg cell. The fine threads represent chromosomes. Move the sperm towards the egg cell until it touches the nucleus of the egg cell. Mix the threads from both lids. This represents the sperm head bursting and the mixing of chromosomes.

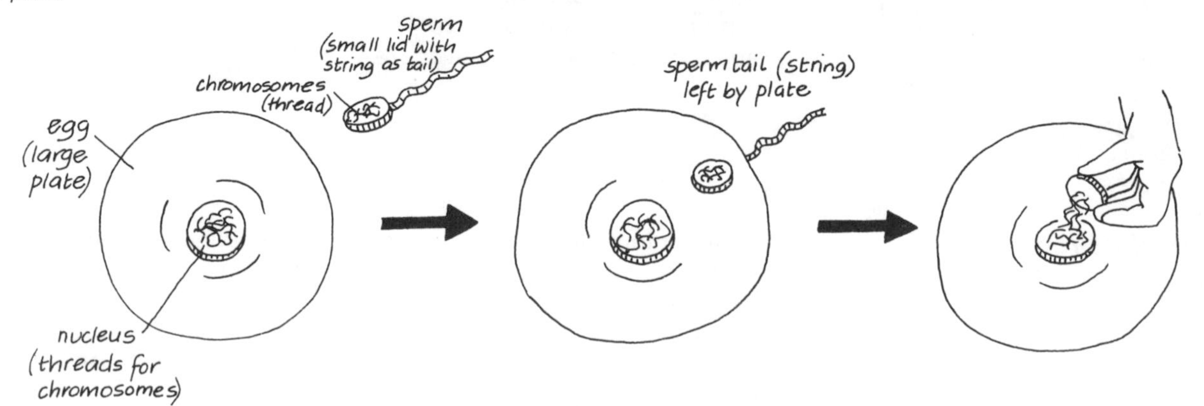

Models of embryo development

At the time of fertilisation the zygote is the size of the smallest dot a pencil can make. (Smaller than a full stop). It is 0.2 mm in diameter. After 2 weeks the embryo is large enough to be represented by clay or Plasticine models. The sizes are as follows:

- 2 weeks old – embryo a ball 1–1.5 mm in diameter
- 3 weeks old – embryo 2.5 mm and umbilical cord formed
- 5 weeks old – embryo still only 5 mm long
- 8 weeks old – now called a foetus 30 mm long and with all human features.

Use toy dolls to explain the functions of the umbilical cord and the placenta.

Protection – the amniotic sac

You will need
- *doll*
- *transparent plastic bag*
- *water*

Place a plastic doll into an empty, clear plastic bag. Pass the doll around the class. Fill the plastic bag with water, place the doll inside and knot the opening so it is sealed. Pass the water-filled bag around and discuss with students what protects a baby inside the mother.

Flower structure

You will need
- *card or plastic*
- *sticks and stones*
- *paper*
- *clay or Plasticine*

Make the major parts of a flower from card or plastic, sticks and clay. Petals can be made from paper.

Ask students to develop the idea.

Look at a variety of flowers, fruit and seeds from the local environment.

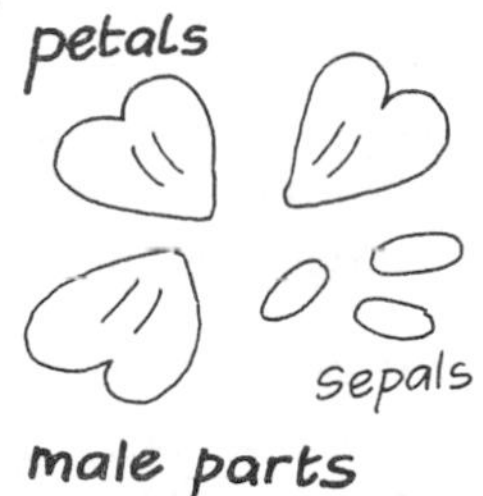

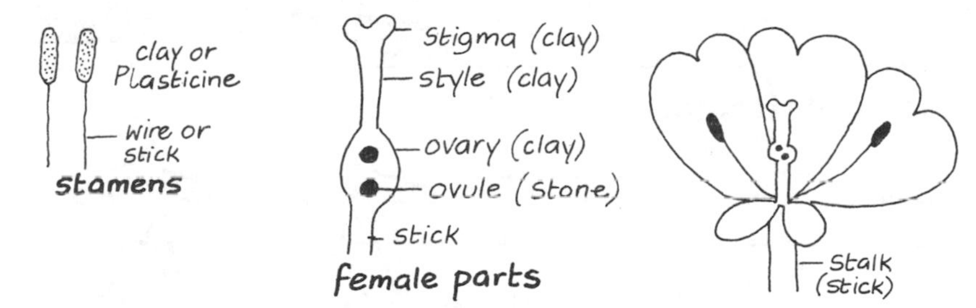

Anthers and pollen

You will need
- *paper*
- *sticks*

Make a model of an anther as shown. Lightly glue small pieces of card or stick onto paper to represent pollen. Alternatively draw circles to represent the pollen. When the paper is folded it represents anthers full of pollen. They are ready to burst open and shed pollen into the wind or onto insects.

Examine the anthers from different flowers with a magnifier (see page 114).

Genetics

- The **nucleus** of every cell (except gametes) contains 2 sets of identical **chromosomes** made of **DNA**. Such cells are **diploid**.
- Sperm and egg cells (**gametes**) contain only one set of chromosomes – they are **haploid**. They are produced by **meiosis**.
- When a sperm **fertilises** the egg the new individual has one set of chromosomes from each parent.
- When cells divide (**mitosis**) the chromosomes divide too and genetic material, the **genes,** combine in different ways. This **recombination** alters the effect of genes.

DNA zip model

You will need
- *a zip*

DNA is wound in a double helix. The strands of the helix are chains of sugars and phosphates. The 2 strands of the helix are linked together by bridges made of pairs of organic nitrogenous bases which are joined to the sugar molecules. A zip provides a good visual analogy.

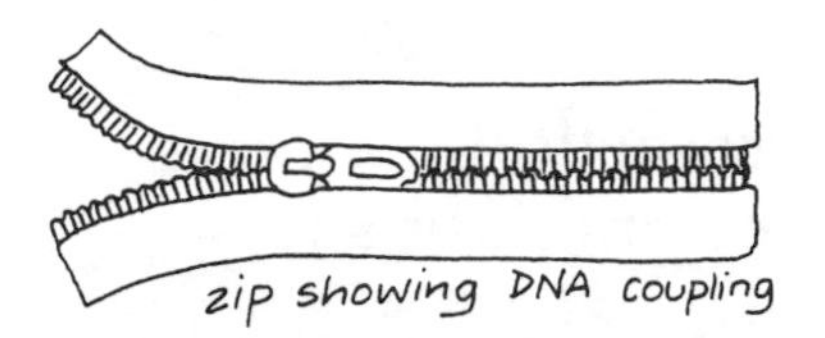

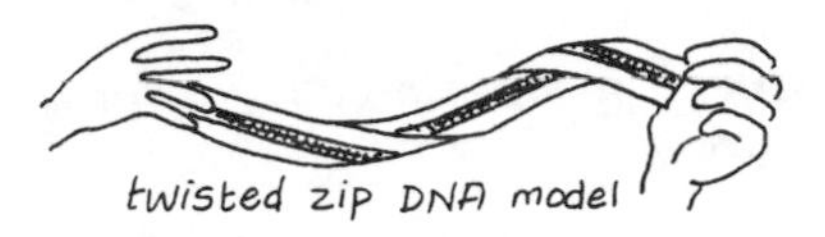

DNA helix model

You will need
- *card or paper strips*
- *4 colours*

A gene can have a sequence of up to 1000 base pairs in a DNA molecule. Make your helix model from strips of strong card or paper. It should be strong enough to twist as shown.

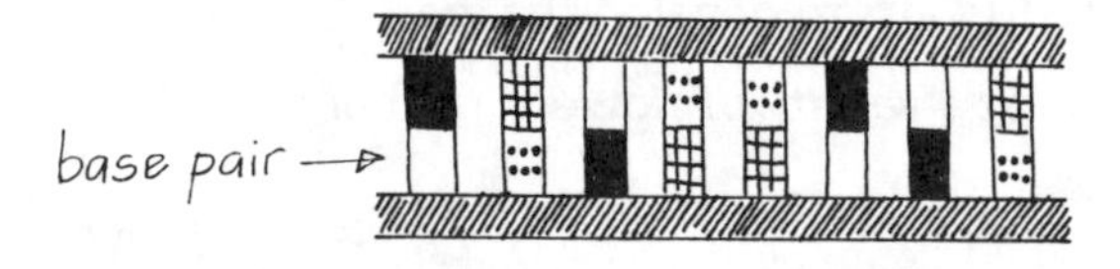

Sex determination

You will need
- *card*

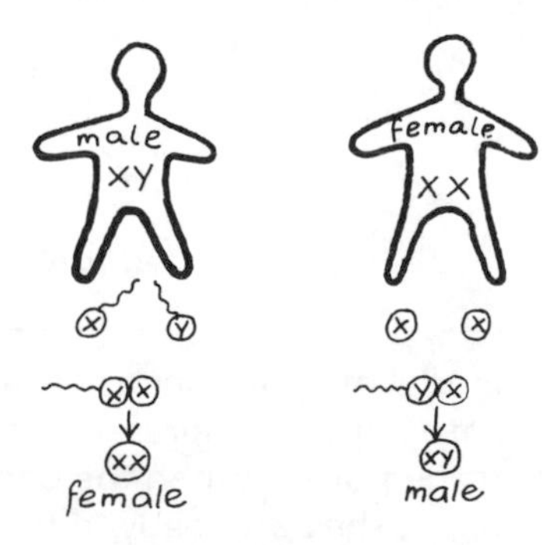

Cut out two shapes, one to represent a male (labelled XY)the other a female (labelled XX). Cut out 4 small circles. Label 3 of them X and label the other Y. These shapes represent the gametes. Move sperm and eggs together to represent fertilisation and sex determination.

Genetic code game

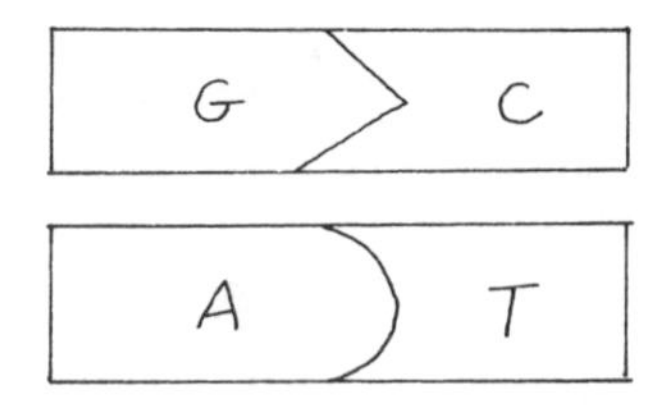

There are always 4 kinds of base in a DNA molecule. These bases always combine in the same pairs: thymine with adenine and cytosine with guanine.

Cut out pieces of card to represent the paired bases. Students must match the bases to 'zip up' the DNA molecule.

This model could be developed to include RNA.

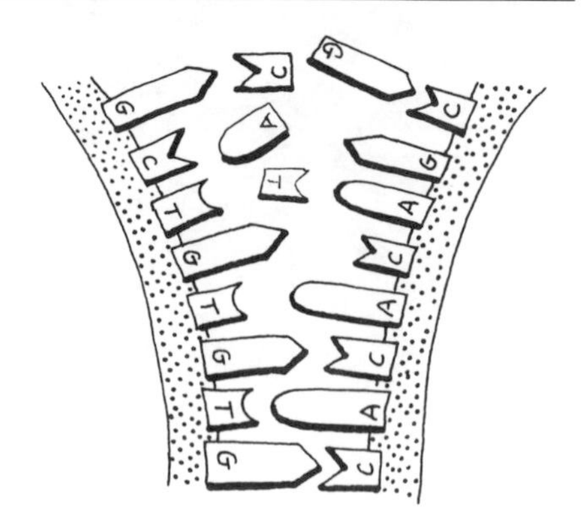

Diploid chromosome set

You will need
- *coloured wool, paper or sticks*
- *large sheet of paper*

Draw the outline of the cell wall and nucleus on paper or a desk top. Make the chromosomes from coloured card, paper, wool or sticks. Make up to 26 pairs of chromosomes (humans have 26 pairs of chromosomes in each cell nucleus). Each pair must be different from all the others, so use different colours, or sizes, to distinguish them. Mix up all the chromosomes, i.e. split all the pairs. Ask students to re-combine the chromosomes to make pairs.

Mitosis – a model

You will need
- *matchsticks or paper strips*

In the model shown here, only one chromosome pair is shown in the original cell. In a human cell, one chromosome from the pair came originally from the sperm, the other from the ovum. 'Parent' and 'daughter' cells have identical chromosomes.

The model would be more realistic and complex if the full complement of 26 pairs of chromosomes were used instead of just one.

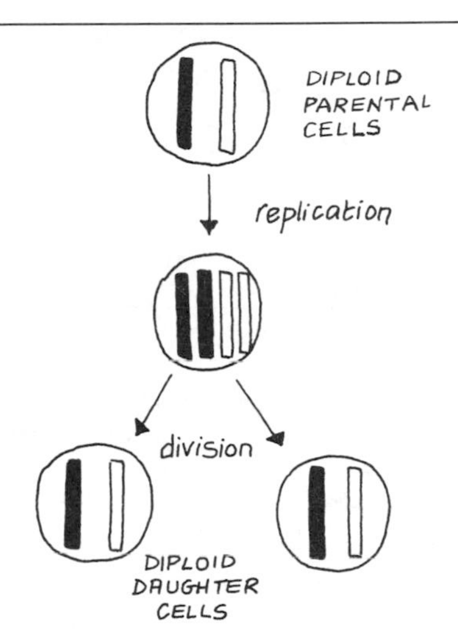

Crossing over – a model

You will need
- *clay or Plasticine*
- *playing cards (optional)*

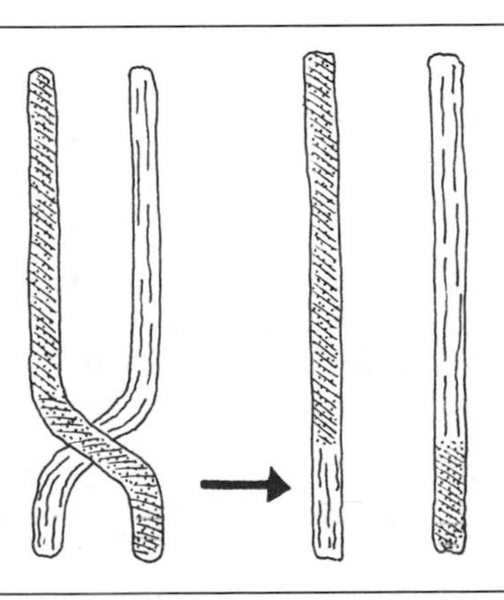

Make 2 chromosomes of different colours in clay or Plasticine. In meiosis pairs of chromosomes come to lie next to each other. At points called chiasmata, parts of chromosomes are exchanged. This crossing over results in exchange of genes.

Ask students how they could show crossing over by using playing cards.

Chromatid models

You will need
- *peg*
- *wire and press stud*
- *paper strips*

During the late stage of prophase in mitosis each chromosome can be seen as 2 parts, called chromatids. These chromatids are joined together by the centromere.

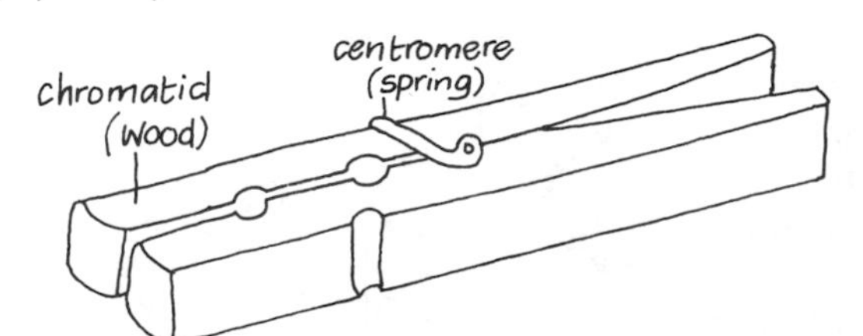

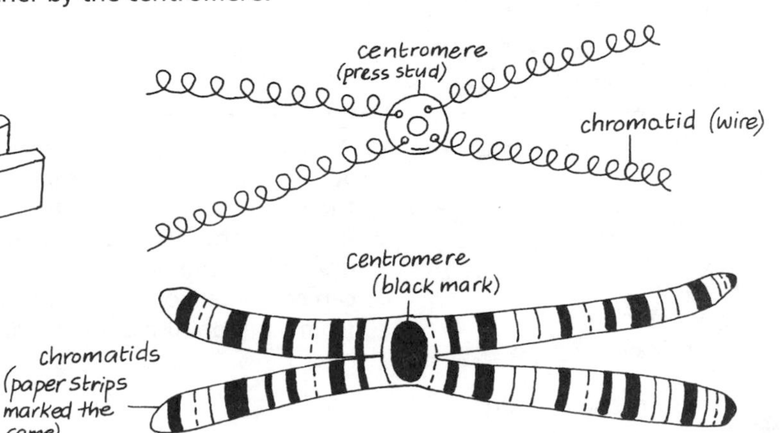

Ecology and ecosystems

- **Ecology** is the study of **organisms** in their **environments**.
- A **community** of organisms, i.e. plants, animals, **micro-organisms**, is an **ecosystem**.
- A **natural ecosystem** is an area not affected by human activities.
- This section gives ideas on how to collect and record organisms in their ecosystems.

Recording and measuring

When you visit a natural ecosystem try not to disturb the animals and plants too much. Write clear records of what you discover on the spot. One recording method is shown here together with useful equipment.

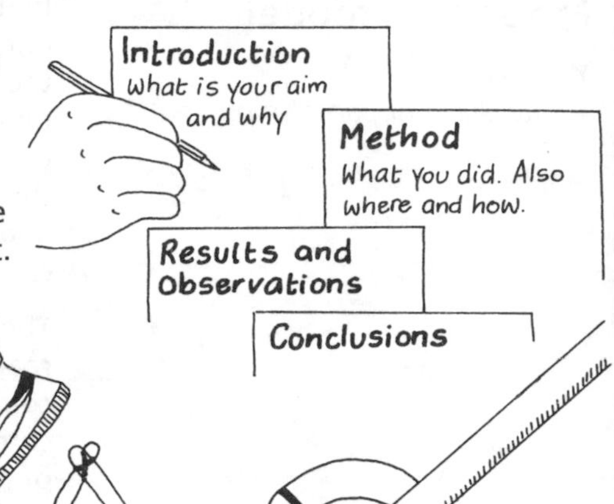

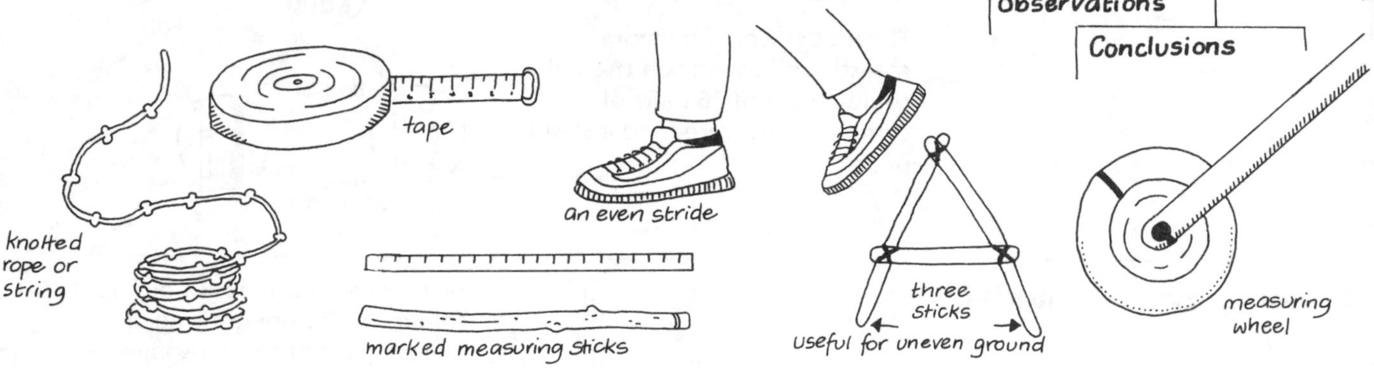

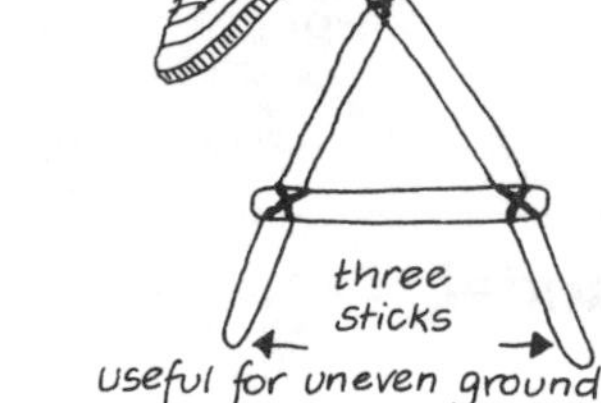

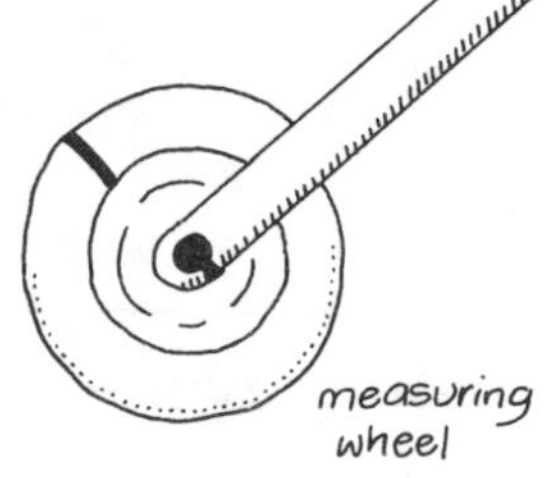

Plant press

You will need
- *old newspapers*
- *heavy weights*
- *2 flat boards*

Collect samples of local plants. Find out the local and scientific names and label the specimens. Place each specimen between pages of newspaper. Place the boards on either side of the newspaper stack and put a heavy weight on top.

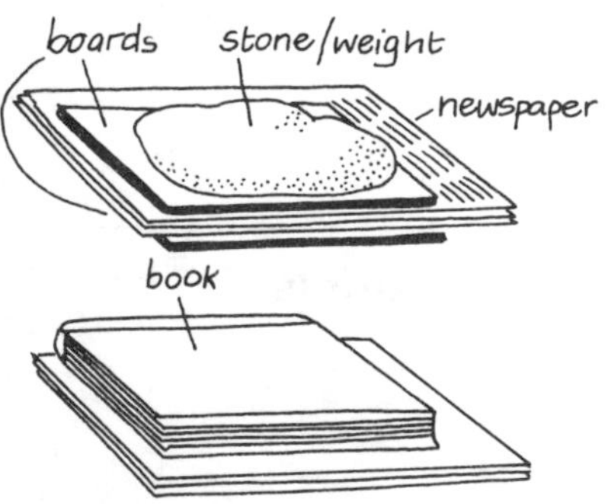

Collecting small insects

You will need
- *small bottle or jar*
- *wide bent tube*
- *rubber tube*
- *gauze or fine cloth*
- *rubber bung with 2 holes*

Place the tube near the insect. Suck in air through the rubber tube and the insect near the mouth of the bent tube will be drawn into the bottle. The tubes can be made from bent straws or from a ballpoint casing bent in heat.

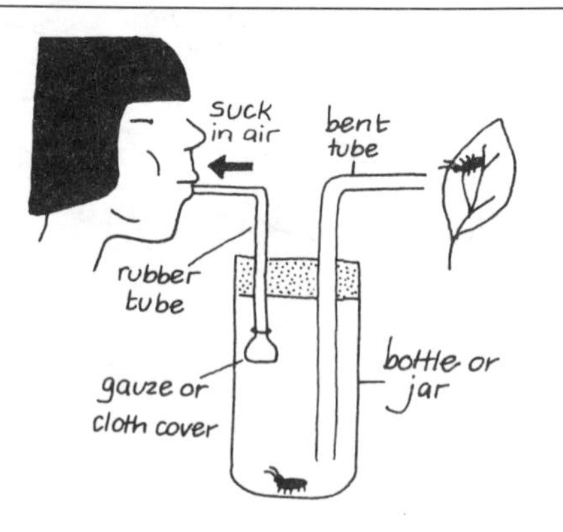

Safety: **Do not allow students to share tubes because of the danger of cross-infection.**

Measuring tree height

You will need
- *a stick or pencil*

Line up the top of the tree with the top of the stick (point 1). Move the fingers to the bottom of the tree (point 2). Turn the stick, keeping the finger in exactly the same place on the stick. Ask a student to walk from the base of the tree until you shout 'stop'. Measure the distance of the student from the base of the tree.

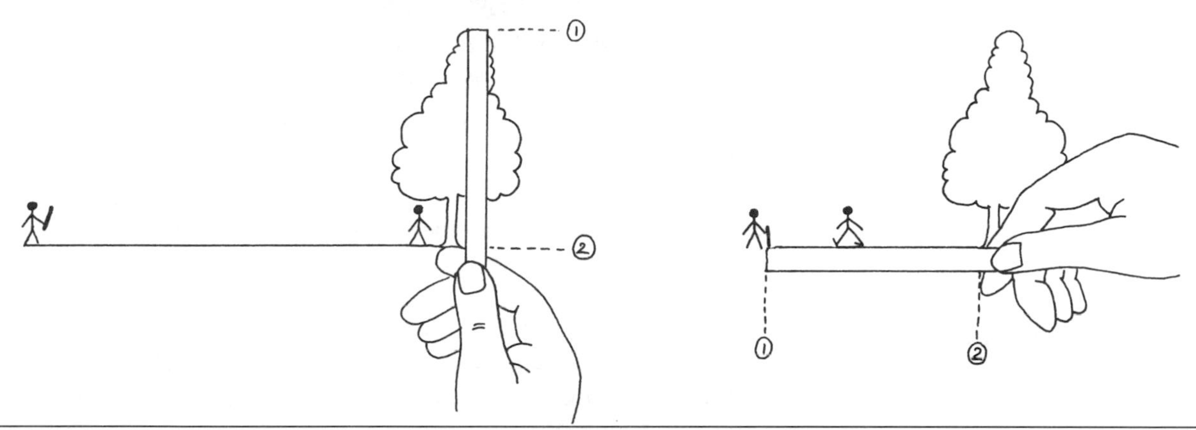

Beating a tree

Put a newspaper or large white cloth under a tree or bush. Beat the branches with a long stick being careful not to cause any damage. The creatures fall onto the paper. Examine and record what you find. If you take creatures away for study return them to where you found them afterwards.

Check nothing dangerous is living in the tree before you beat it or there may be a few surprises! Do not damage the tree or bush.

Line transect survey

You will need
- *long piece of string*
- *2 pegs or heavy stones*

Stretch a length of string between 2 pegs. Make a list of all species of plant which touch the string or lie under it. Make an elevation drawing (to scale) of each line transect. Compare transects in different locations.

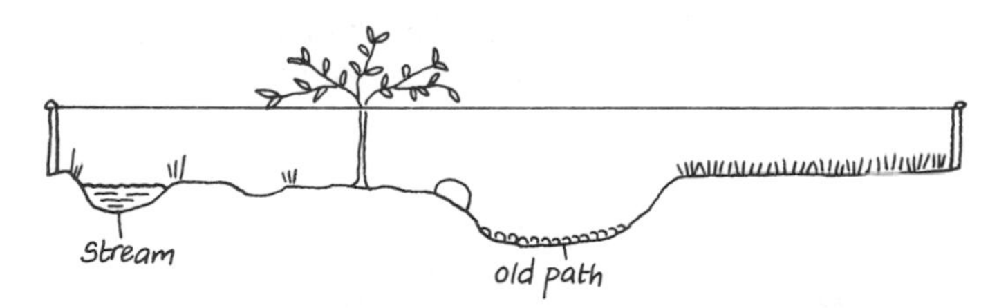

Sampling squares

You will need
- *cardboard or wood*
- *wire*
- *rubber tube (optional)*

Squares can be made in various ways and various sizes. A square of 1 metre works well. After throwing or placing the square on the ground, record the names or descriptions of organisms found in the square. Use the square to sample species in different locations around the school.

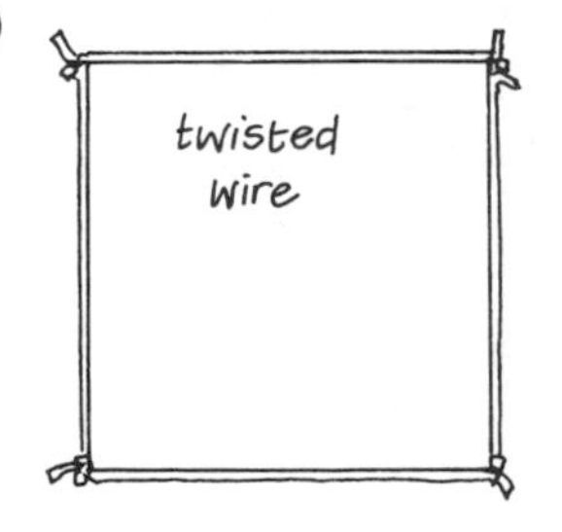

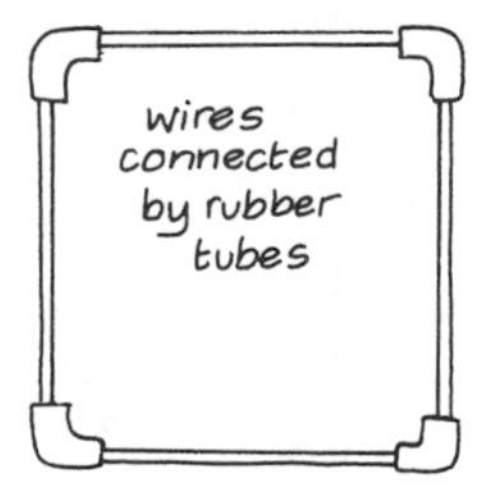

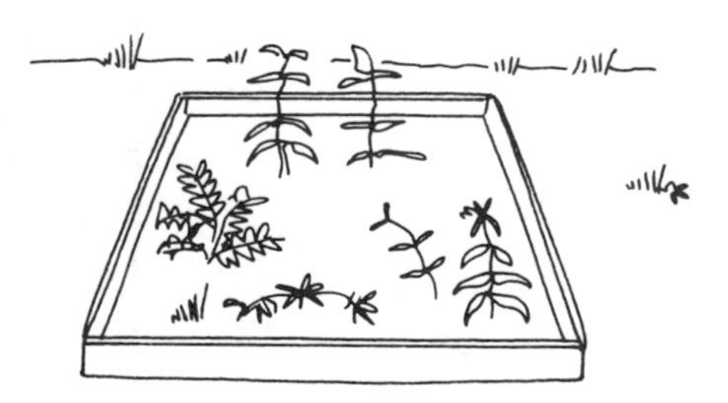

56 The balance of nature

- All living things ultimately depend on the Sun.
- **Producers** (plants) trap energy from the Sun during **photosynthesis** (see page 38). Producers are eaten by **primary consumers (herbivores)** and these may be eaten by **secondary consumers (carnivores)** which may in turn be eaten by **tertiary consumers (other carnivores)**.
- The complex relationship between producers and consumers can be represented in **food chains**, or **food webs**.
- Animals and plants living in an **environment** often show **adaptations** or behaviour patterns which help them survive.
- In the natural world all matter is constantly recycled, but human activities are upsetting the **balance of nature** and the long-term consequences may be devastating.

Food chain mobile

You will need
- *stiff cardboard*
- *scissors*

Cut links of the food chain from stiff cardboard. Label each link with one part of the food chain. Put the links together to make a chain.

Make both simple and more complicated chains.

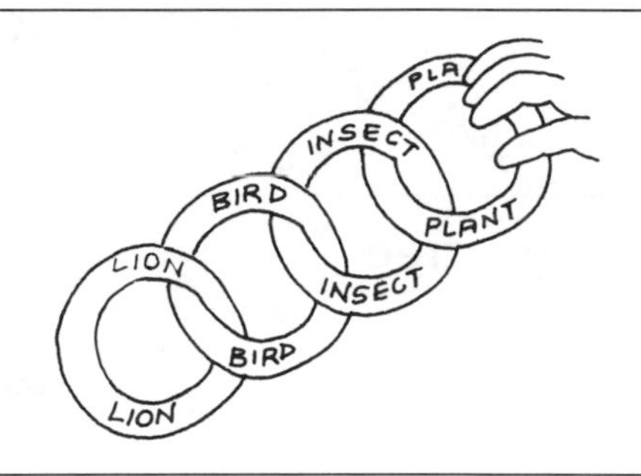

Food webs

You will need
- *card*
- *pictures of animals and plants (optional)*

Either draw pictures of animals and plants on cards or stick on pictures cut out from magazines etc. Make arrows and write on them the links shown. Arrange the cards and arrows to make a food web.

This could be extended into a card game for several players.

Carbon cycle cards

You will need
- *card*
- *paper strips*

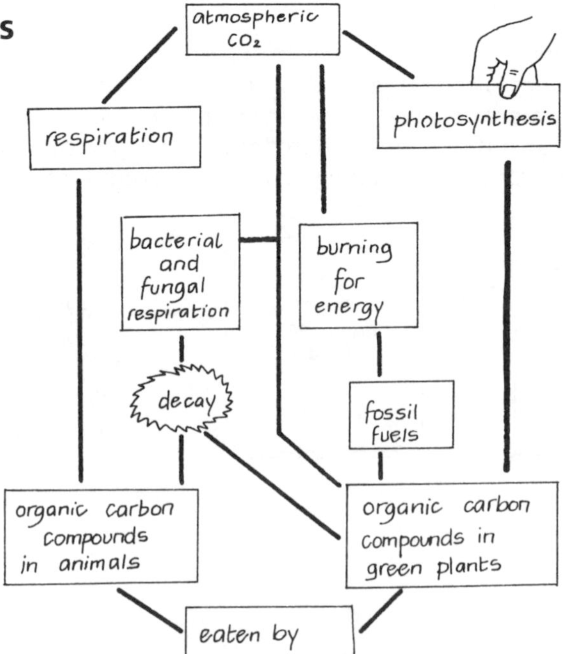

Cut out cards showing stages of the carbon cycle. Link them together with the paper strips to make a balanced carbon cycle.

Discuss with students the consequences of increasing one stage, e.g. burning extra fossil fuels.

Cards could be made for other cycles too.

Camouflage

Disappearing moths

You will need
- *newspaper*
- *white paper*

Cut moth shapes from newspaper and white paper. Place both types of moth onto newspaper, and then onto white paper. Note which moths are easier to see.

Camouflage and protection

You will need
- *a long piece of string*
- *4 pegs*
- *matchsticks*

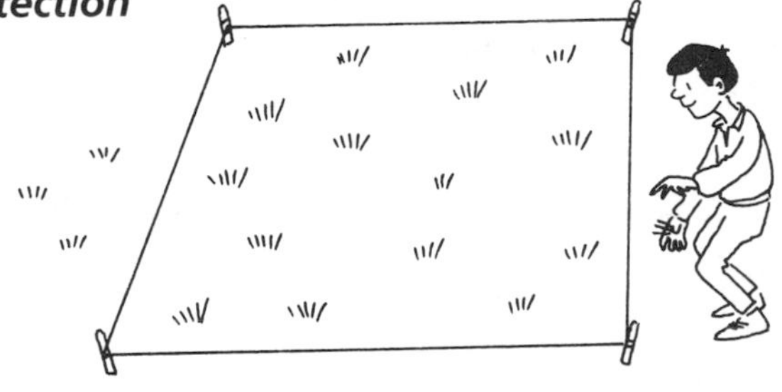

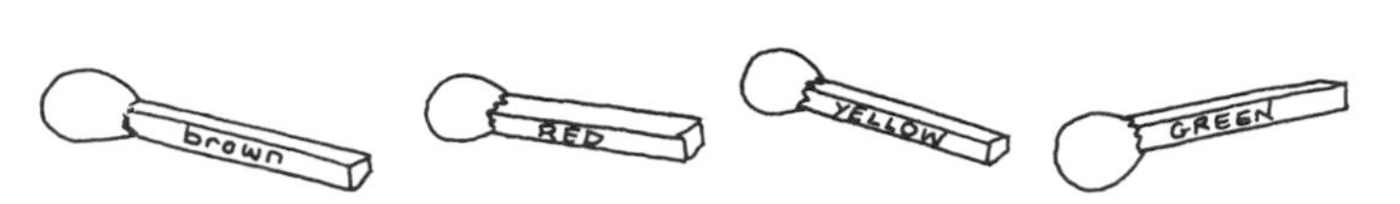

Mark out an area of grass with the string and pegs. Colour the matchsticks with different inks or paints. Make some the same colour as the grass and others in very bright colours. Drop the matches over the area of grass. Which matches are easiest to find?

Discuss with students why camouflage would be an advantage to a small maggot and why it would help a predator too.

Survival behaviour

You will need
- *2 glass plates or petri dishes*
- *maggots*
- *dry and damp paper*
- *black paint*

Maggots die if left in hot sunshine because they dehydrate. You can demonstrate how behaviour aids survival.

Reactions to light

Paint or cover one half of each of the plates. Put the plates together so that half is dark and half in bright light. Put 10 maggots into the centre of the bottom plate and put the 'lid' back. Count how many maggots are in each side every 10 minutes.

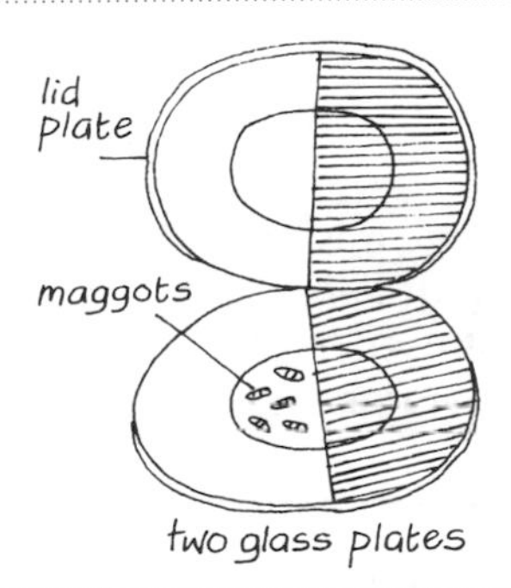

Reactions to humidity

Put dry filter paper on one side of a plate and damp paper on the other. Put a plate on top and cover it with a cloth so it is dark underneath. Count how many maggots are on each side every 10 minutes.

Investigate several conditions at once. For example, put damp filter paper on one half of the 2 plates. Is the result the same if both plates are in sunlight? Which is more important, dampness or darkness, i.e. do maggots prefer light and damp or dark and dry?

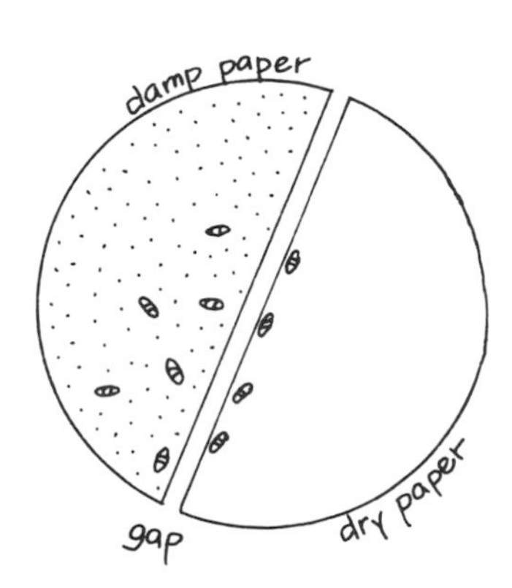

58 Health matters

- **Diseases** caused by micro-organisms (**viruses** and **bacteria**) spread by close contact, e.g. HIV, or through the air or water, e.g. 'flu and coughs.
- Some diseases are not caused by **micro-organisms**, but by unhealthy habits such as smoking.
- Severe diarrhoea is often fatal to babies because it causes **dehydration** and loss of **electrolytes**.
- HIV is a virus which, in time, leads to AIDS.

Coughs and sneezes spread diseases

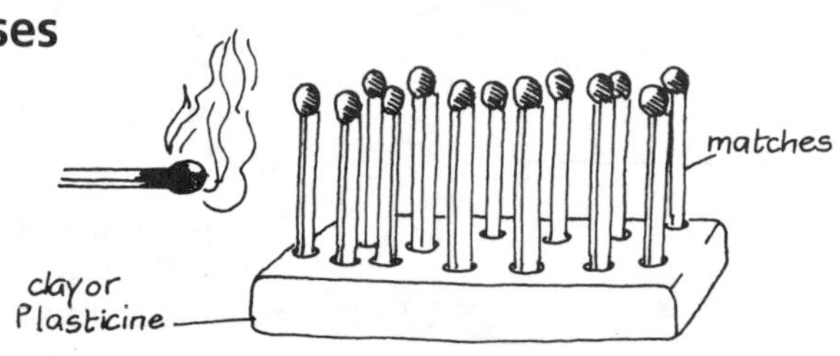

Smoking and health

You will need
- *2 syringes*
- *filter or newspaper*
- *cigarette*

Remove the needle end from one syringe (syringe 2). Remove the plunger from the other syringe (syringe 1) and make a larger hole in the needle end. Join the syringes as shown. Place a piece of filter paper or newspaper between the 2 syringes. Place the cigarette in syringe 1 and light it. Draw air through the cigarette several times and you will see a dark stain spreading across the filter paper. This is tar from the cigarette.

Ask students what happens to the tar if a person smokes the cigarette and discuss its effect on health.

Safety: **It is essential any syringes are sterile when first used in this apparatus.**

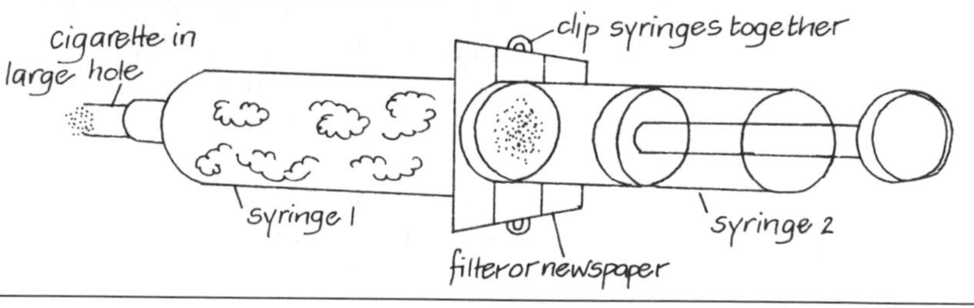

The dangers of dehydration

Water baby

You will need
- *plastic bottle, can or gourd*
- *2 corks*
- *water*

Make a model baby from the bottle, can or gourd. The hole in the top represents the mouth. Put 2 holes in the bottom of the container. Make a small hole to represent water loss through urine. Make a large hole to represent the anus. Put corks in both these holes. Fill the 'baby' with water.

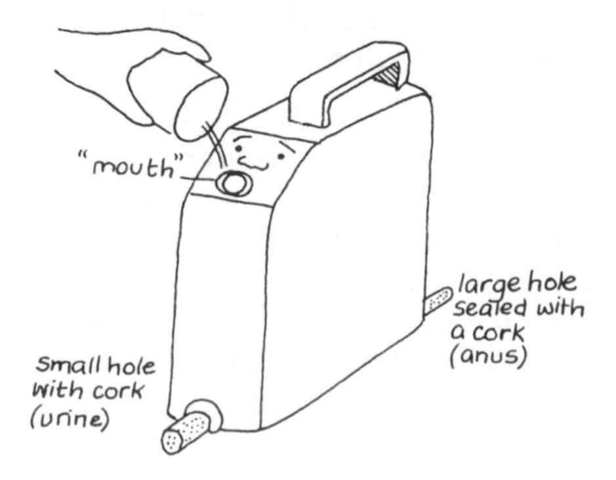

Remove the smaller plug and water will be lost slowly. However, diarrhoea can cause severe loss of water, as removing the larger plug illustrates. Water lost through the holes can only be replaced through the 'mouth'. If more water is lost than is taken in dehydration occurs and this can be fatal, especially in small babies.

The ORS miracle

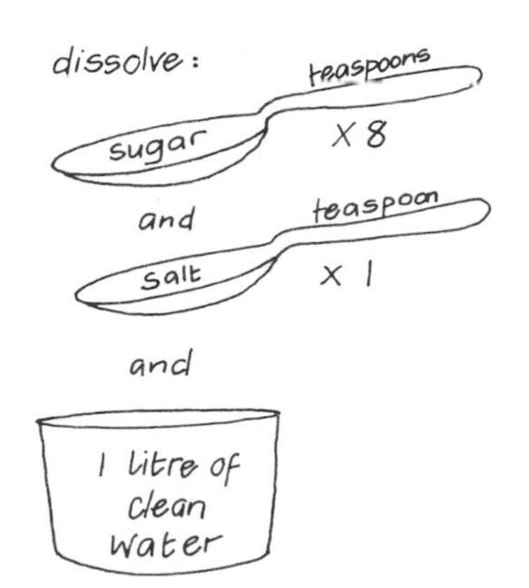

Dehydration caused by diarrhoea is probably the biggest killer of children in today's world. Our bodies need water to function normally, but we also need a particular concentration of essential electrolytes, e.g. sodium and potassium. These electrolytes are lost in diarrhoea and they must be replaced. Drinking water alone will not save the life of a person who is dehydrated and has lost too many electrolytes. To replace some essential electrolytes and water, the baby, or adult, should gently be made to drink the Oral Rehydration Solution (ORS) shown here.

***Note:* This is an emergency solution and does not contain all electrolytes. A severely dehydrated baby may need a more complex solution if diarrhoea persists.**

The aids epidemic

HIV acting

You will need

- *cards*
- *pins or tape to attach cards*

Make cards to attach to students. They should contain a mixture of the following – HIV; diseases, e.g. TB, diarrhoea; white blood cell. One of the pupils should represent the human body. Several 'white blood cells' should be protecting one 'body' to begin with. Ask students to act out the spread of HIV.

White blood cells protect the body from diseases. HIV knocks out the white blood cells and so they can no longer protect the body. This leaves the body open to attack by germs of all kinds. Eventually the body is overcome by diseases which are normally not fatal.

Passing on HIV

You will need

- *cards*
- *starch solution*
- *iodine solution*

On the cards write down some sexual case histories. Give each student a card at random. The owner of the card is to follow the behaviour indicated on the card, e.g. faithful to one partner, many partners, no partners. Give a few of the students a cup of starch solution and, give all the others a cup of water. Ask students to follow the case history of the cards and to mix the contents of their cups when they have a partner – mixing represents sexual contact. At some point 'HIV test' the contents of the cups using a few drops of iodine solution. If the solution goes dark then it means there is starch (representing HIV) in the cup.

Discuss how fast the virus spreads. Also discuss how its spread could be prevented or slowed down.

Could you develop this idea further?

Raw materials

- The **atoms** or **molecules** of raw materials are processed to make new substances.
- New substances can be made by **combining** two **elements** or by breaking down **compounds**.
- Some of the key processes are shown here: **electrolysis, fractional distillation, cracking** and **polymerisation**.

Electrolysis

During electrolysis electricity passes through a liquid and decomposes it.

Electrodes

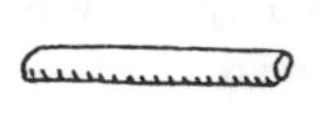

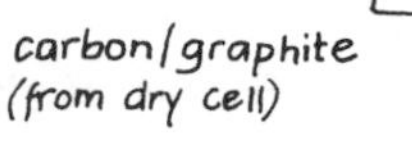

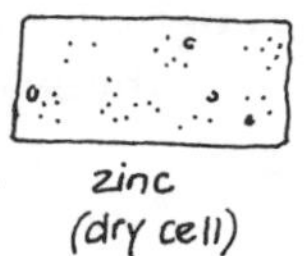

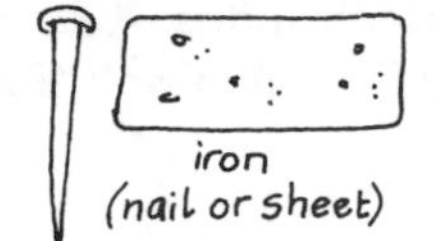

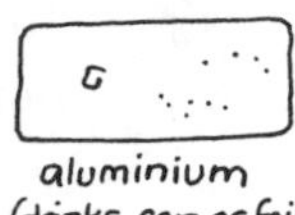

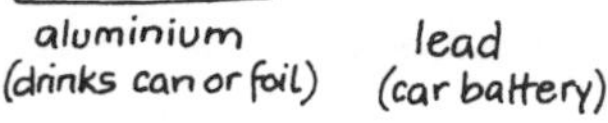

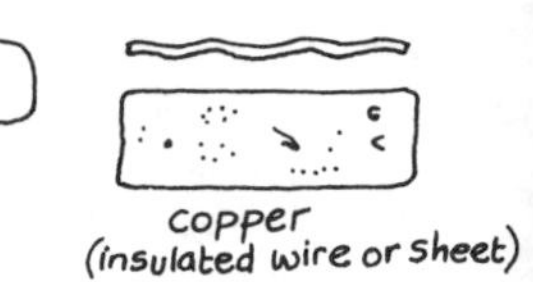

Electrolytes and electrodes

electrolytes	cathode	anode
HCl (half-conc)	graphite	graphite
alkali – chlorides (2M)	iron / graphite	graphite
alkali – hydroxide (2M)	iron	iron
H_2SO_4 (2M)	lead or graphite	lead
Na_2SO_4 (16%)	iron	lead
$CuSO_4$	graphite	graphite or lead

Electrode holders

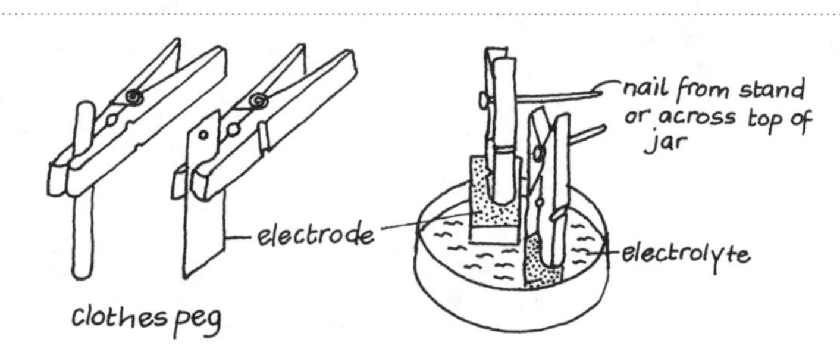

Safety goggles should be worn.

Electrodes

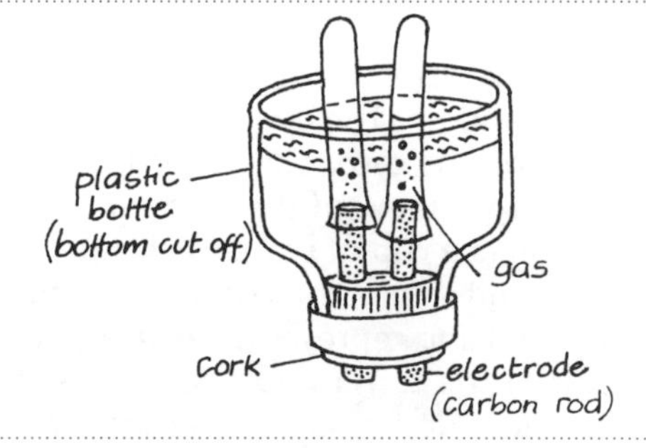

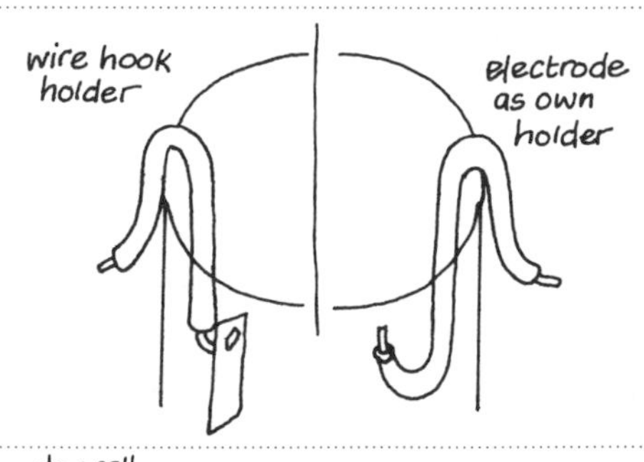

Electroplating

You will need

- *clothes peg*
- *copper strip*
- *iron nail*
- *copper(II) sulphate solution*
- *dry cell*
- *safety goggles*

Make sure the nail is the cathode – it must be wired up to the negative terminal of the dry cell. The copper strip then becomes the anode. After current has flowed for a time the nail becomes pinkish as copper is deposited on it. The nail is being electroplated with copper.

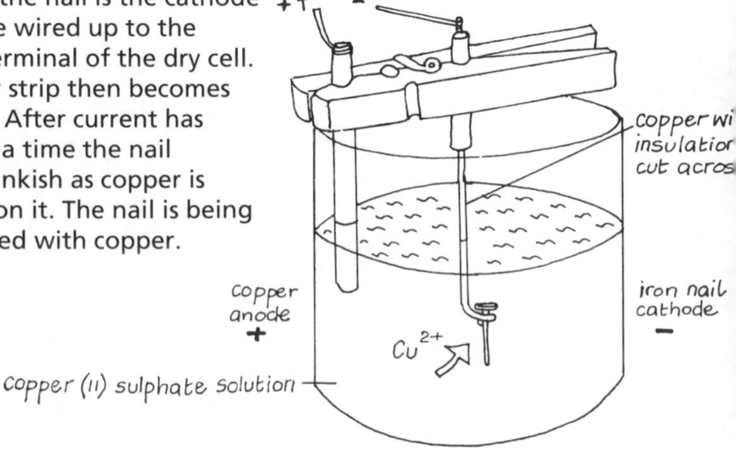

Refining oils

Fractional distillation of crude oil

You will need
- *2 test tubes*
- *rubber bung*
- *plastic tube*
- *safety goggles*

Care is needed with inflammable substances.

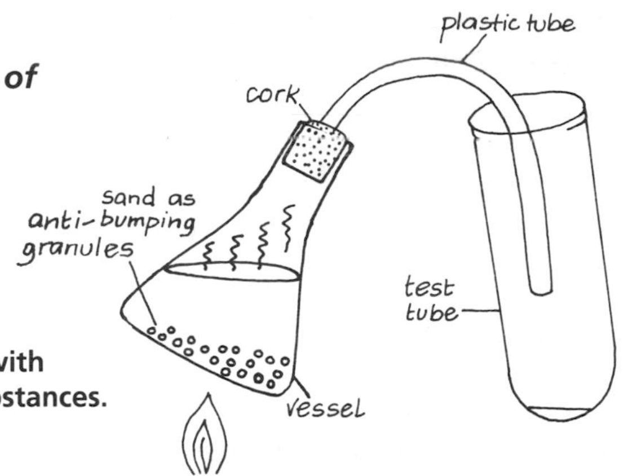

Put the crude oil into one test tube. Add sand and then seal the tube with the bung and plastic tube. When the crude oil is heated vapours will come off the various fractions. With careful control of temperature the fractions can be separated.

If crude oil is unavailable then the following is a successful mixture:

1 part diesel
1 part kerosene
1 part duplicator spirit
0.5 parts sump oil (for colour)

Cracking

You will need
- *boiling tube*
- *broken pottery*
- *bung with plastic tube*
- *plastic bottle*
- *large dish of water*
- *safety goggles*

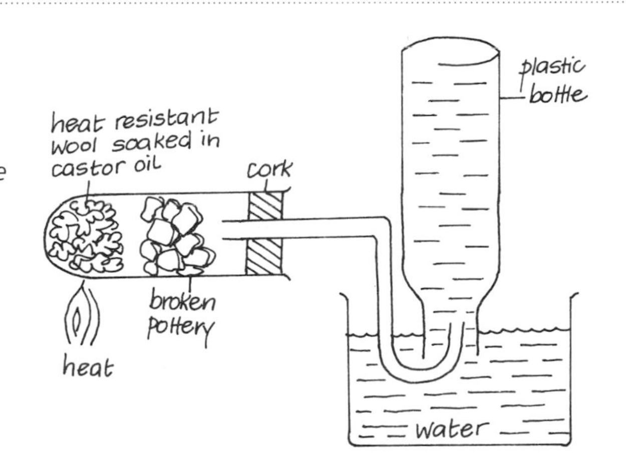

Set up the apparatus as shown. Cracking is the process by which large hydrocarbon molecules are broken down into smaller ones. Castor oil 'cracks' over a broken tea cup or flower pot into a gas which can be collected in the plastic bottle. The gas burns well.

Cracking needs a lot of heat. A gas burner may be best. Do not use test tubes or light bulbs. It is essential to use a hard-ened glass boiling tube.

Polymerisation

This is the process where many individual small molecules link together by covalent bonds to form larger molecules. Some well known polymer substances are illustrated below.

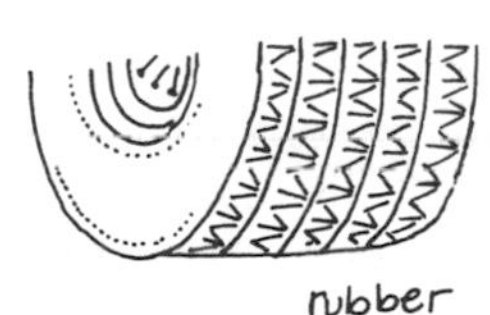

Acting out polymerisation

The small molecule is ethene which polymerises to poly(ethene) – often called polythene.

double bond

H H C

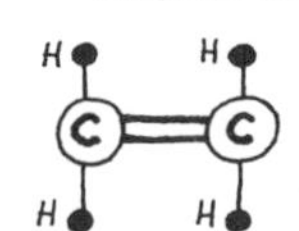

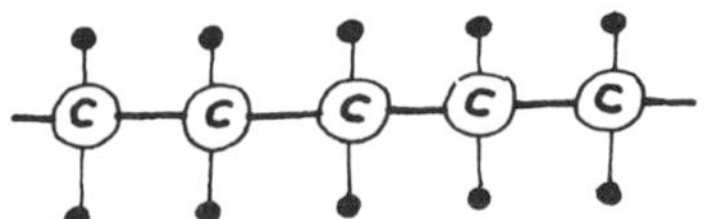

<u>62</u> Separating mixtures

- A **mixture** consists of 2 or more substances which are not chemically bound together – the substances can be separated from each other.
- **Pure** substances contain only one type of **molecule** (or **atom** in the case of pure **elements**). Some of the techniques used to separate the substances in a mixture are shown here.
- Separation techniques are often used to **purify** one of the substances in a mixture.

Filtration

You will need
- *glass container*
- *filter funnel and filter paper OR*
- *cloth and wire support*

Set up the apparatus as shown. Investigate which types of cloth and paper work best. Try colourless newspaper (i.e. a part with no printing in on it), paper towel and toilet tissue.

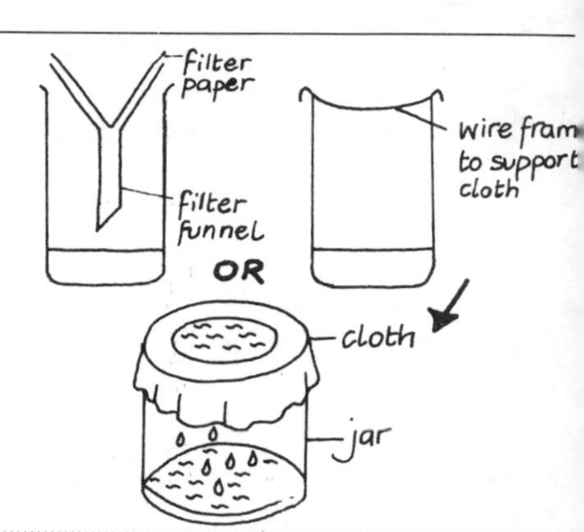

Using filtration

Ginger beer is made using filtration.

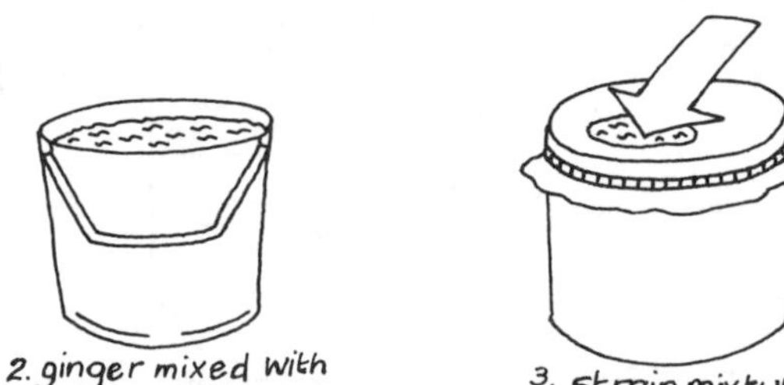

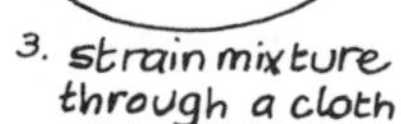

Separating solids

Sand and salt

You will need
- *a mixture of salt and sand*
- *container*
- *filtering apparatus*
- *bottletop and support*
- *heat*

If one of the 2 solids dissolves in water and the other does not, the solids can be separated quite easily, e.g. salt can be separated from sand. First water is added to the mixture – the salt dissolves. The sandy liquid is then filtered. The filtrate is then evaporated to leave behind pure salt. A small amount of filtrate can be evaporated as shown.

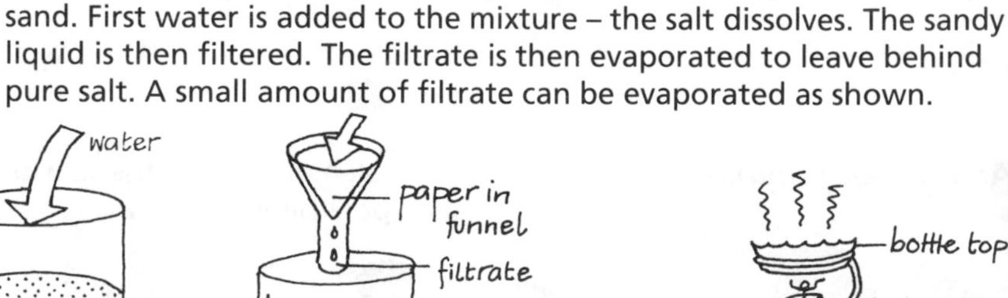

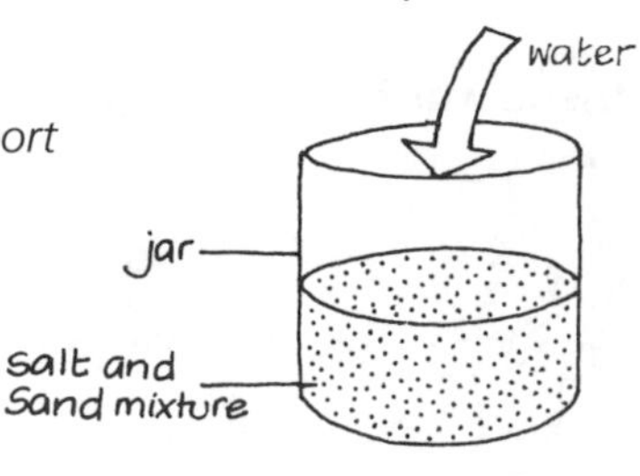

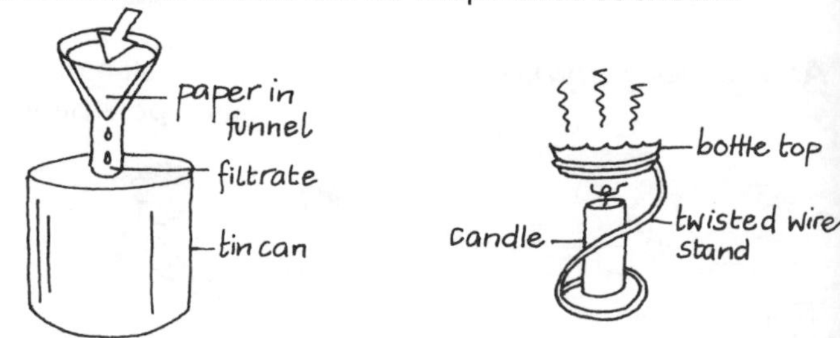

Iron and sulphur

You will need
- *iron and sulphur mixture*
- *magnet*

Cover the magnet with paper. The magnet will attract only the iron, leaving sulphur behind.

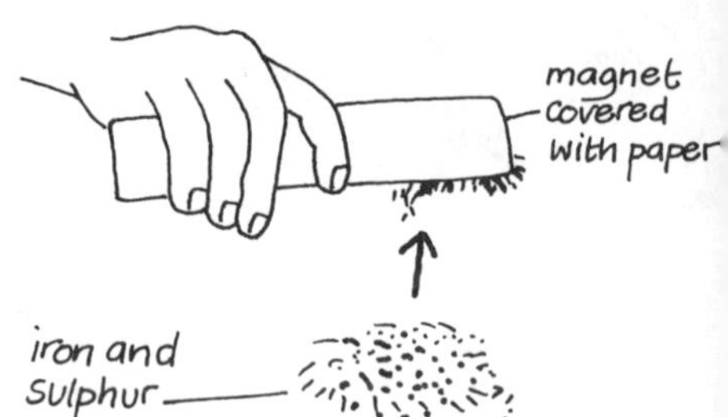

Mothballs and salt

You will need

- *mixture of mothballs (naphthalene) and salt*
- *glass funnel or tumbler*
- *tin can*
- *heat*
- *safety goggles*

If the mixture is heated the naphthalene sublimes (turns into a gas) and rises into the funnel. It condenses out on the inside of the funnel, leaving the salt in the tin.

Iodine can be used instead of naphthalene.

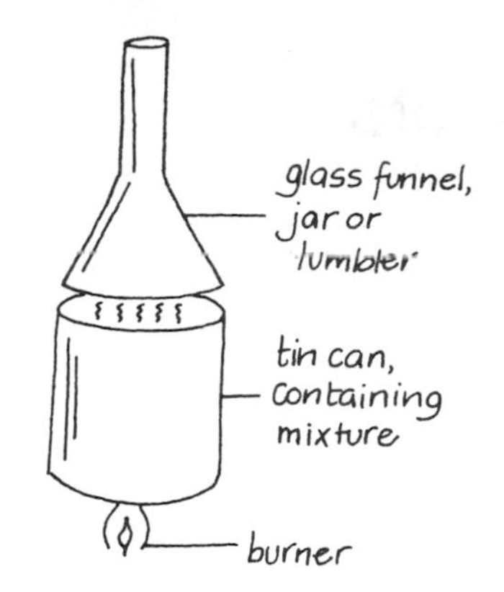

Care is needed to avoid inhaling fumes. Naphthalene is also very flammable.

Separating liquids

Immiscible liquids

You will need

- *mixture of kerosene and water*
- *plastic bottle*
- *cork*
- *rubber tube*
- *peg or clip*
- *safety goggles*

When 2 liquids will not mix with each other they are said to be immiscible. One liquid will sink below the other and can be drawn off as shown.

Other suitable mixtures are: groundnut oil and water; palm oil and water; petrol/diesel and water; castor oil and water. Palm oil is particularly effective because it is brightly coloured.

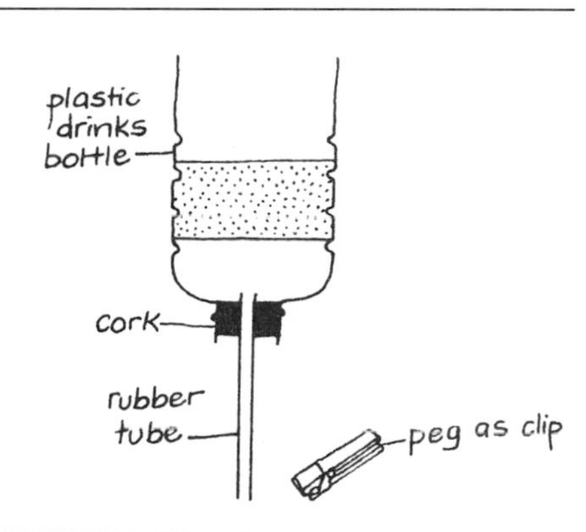

Using the technique

Palm oil is extracted as shown. The final process depends on the immiscibility of palm oil and water. The oil floats to the surface and is skimmed off.

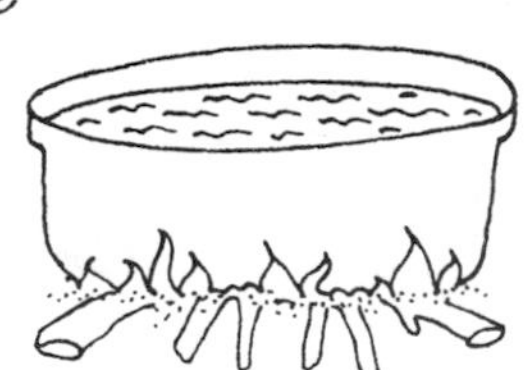

Separating other mixtures

If 2 liquids will mix they are said to be miscible. Fractional distillation (see page 60) is used to separate each of the 3 mixtures shown.

Ask students to separate a mixture of alcohol, water and crude oil.

Clean water

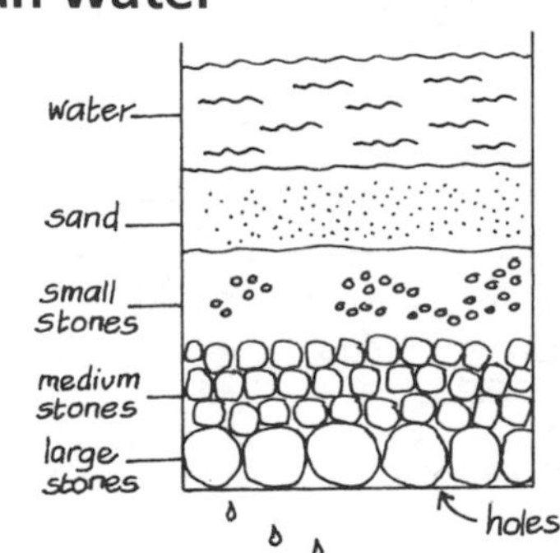

Water can be purified using separation techniques. Untreated water contains both soluble and insoluble matter and also bacteria which can all cause harm.

Filtering removes the insoluble matter such as sand and soil. If the filter is very fine it will also remove bacteria and other micro-organisms.

Distillation removes all insoluble matter and also kills bacteria.

Boiling kills bacteria and other micro-organisms.

Metals

- All **elements** are either **metals** or **non-metals**.
- Metals have particular chemical and physical **properties**, e.g. they are all solid at room temperature except mercury and many of them **react** with acid to form hydrogen.
- Metals can be placed in a sequence according to their reactivity – called the **reactivity series**.
- Iron is one of the most widely used metals, but it **oxidises** in damp conditions – it corrodes or rusts.

Properties of metals

Malleability

You will need
- *hammer or large stone*
- *samples of metals*

Hammer the metals.

Discuss the meaning of the word malleable and the effect of heat and cold on different metals and their malleability.

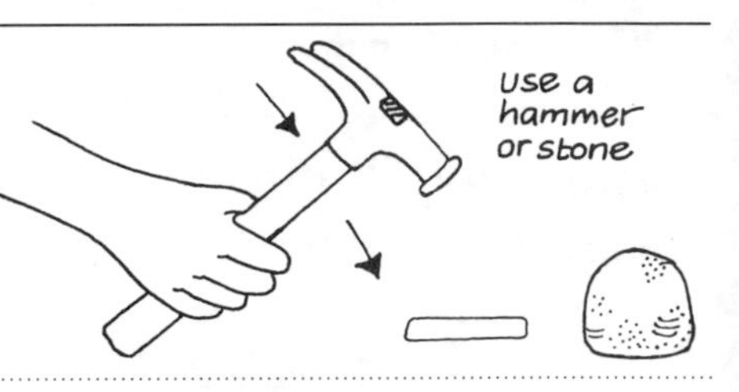

Ductility

You will need
- *supports*
- *metal wire*
- *weights*

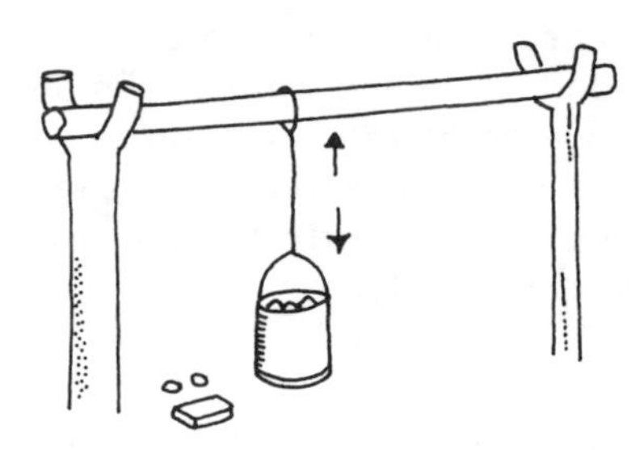

Suspend the wire between supports and hang a weight onto the free end. One method is illustrated. Measure the length of the wire. Add weights and the wire will stretch.

As an extension, compare the ductility of wires made from different metals.

Conductivity

You will need
- *dry cell*
- *wire*
- *light bulb*
- *metal samples*
- *non-metal samples*

Set up the circuit shown. All metals will conduct electricity.

Investigate if any non-metals conduct electricity.

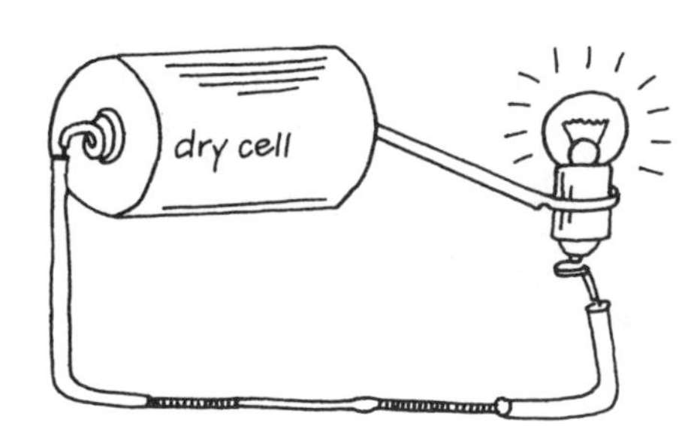

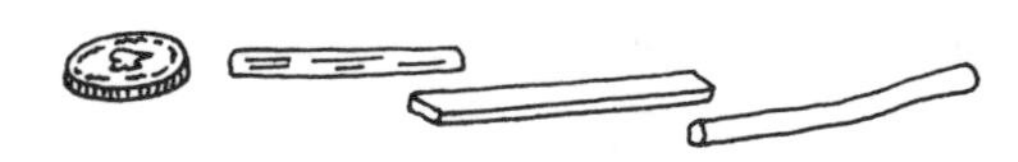

Reactivity series of metals

You will need
- *glass sheet*
- *large sheet of paper*
- *metals (small pieces which must be clean), e.g. magnesium ribbon, zinc granules, lead shot or sheet, copper sheet, iron filings*
- *solutions containing metal ions, e.g.copper(II) sulphate – Cu^{2+}; iron(II) sulphate – Fe^{2+}; magnesium sulphate – Mg^{2+}; zinc sulphate – Zn^{2+}; lead nitrate – Pb^{2+}*

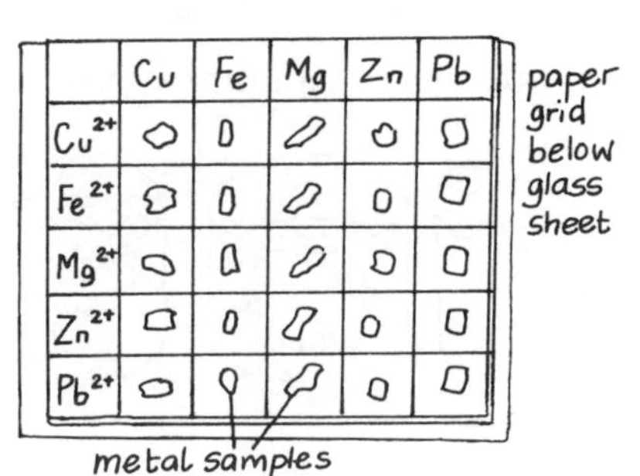

Mark the sheet of paper to make a grid as shown. Place the glass sheet over the paper grid. Place the metals on the right squares. Add 2–3 drops of one of the solutions to each of the metals and observe any change. If a black coating forms on the metal it indicates that the metal ions are being displaced from the solution and deposited onto the metal.

Growing metal crystals

You will need

- *lead acetate (ethanoate)*
- *rainwater*
- *zinc strip*
- *glass container*
- *stick or nail*

Add 4 spatulas of lead acetate (powder), or lead nitrate powder to half a cup of water. Suspend the strip of zinc in the lead acetate solution. After a few days crystals grow on the plate.

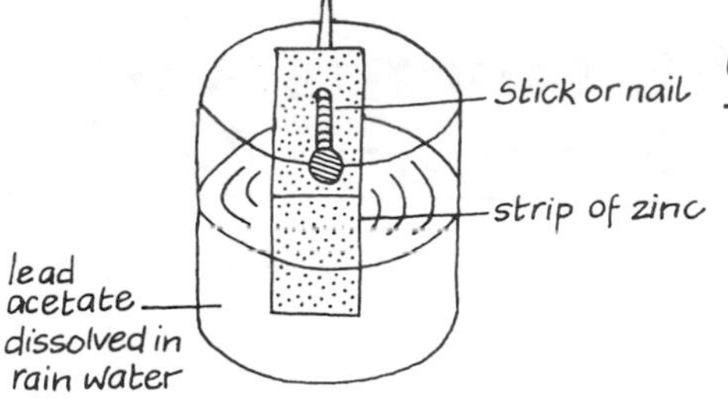

Corrosion (rusting)

Rusty nails

You will need

- *3 containers*
- *3 iron nails*
- *bung/cork*
- *boiled water*
- *tap water*

Place a nail in each of the containers and leave for a day. The only nail which does not rust is the one in the sealed jar of boiled water. Boiling the water removes the oxygen and sealing it prevents oxygen from the air dissolving in it.

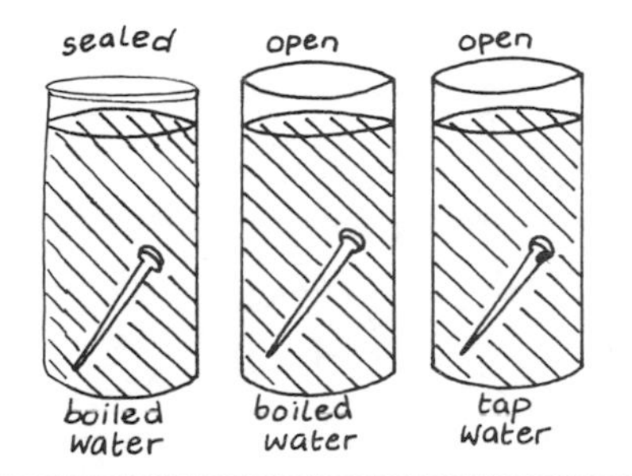

Oil-covered nails

You will need

- *3 nails*
- *3 containers*
- *oil*
- *salt water*
- *tap water*
- *sand and water*

Cover all the nails with oil and place one in each of the containers. Leave them for a day. Note whether the oil has protected the nails from rusting. Shake up the containers, so the sand rubs against the nail and discuss the effect this has on the speed of rusting.

Investigating rusting

Tin cans

You will need

- *old tin can*
- *oil*

Make 2 large scratches on the surface of a tin can (not an aluminium can often used for drinks). Put a thin layer of oil onto one scratch. Leave the tin exposed to the air for a few days. Note which scratch rusts.

Steel wool

You will need

- *steel wool*
- *candle*
- *2 glass containers*

Wet the steel wool and place some in each container. Seal one container. Place a lighted candle in the other container. When the candle has burnt for several minutes seal the container with a lid. The candle will go out eventually. Leave both containers for 2 days. The steel wool in the container with the candle should not rust as much because oxygen has been removed by the candle.

Discuss with students what factors are needed for rusting to take place. Also discuss how rusting can be prevented.

Preventing corrosion

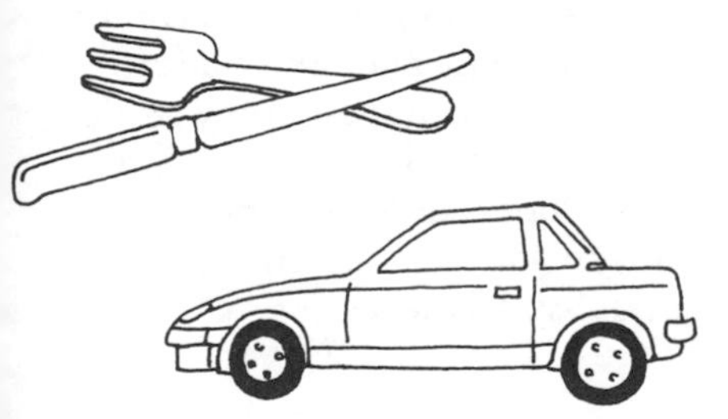

Rusting of iron can be prevented in various ways.

Stainless steel is an alloy of iron and chromium which does not rust or stain.

Paint protects metal, e.g. car paintwork protects the bodywork from corrosion.

Oil, like paint, prevents water and oxygen reaching the surface of the metal. Oil also lubricates machine parts and so prevents scratching and wear.

Elements and compounds

- **Elements** contain only one type of **atom**.
- **Compounds** are made of 2 or more different types of atom.
- Atoms consist of **protons, electrons** and **neutrons**.
- Electrons are arranged in **electron shells** round the **nucleus**.
- The particles of a substance are held together by attractive forces called **bonds**.
- The arrangement of **particles** is different in different materials, e.g. crystal particles are arranged in a **lattice**.

Models of atoms

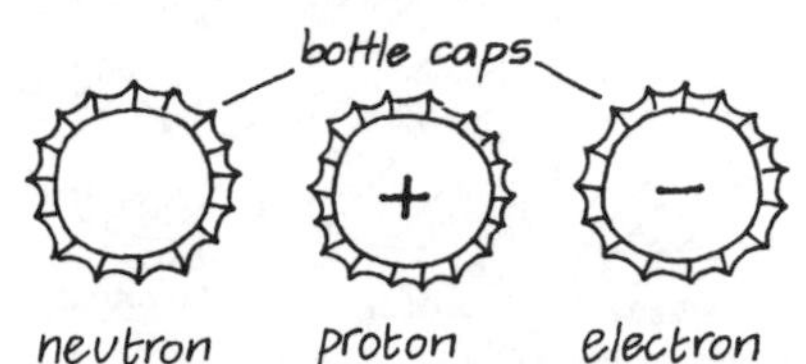

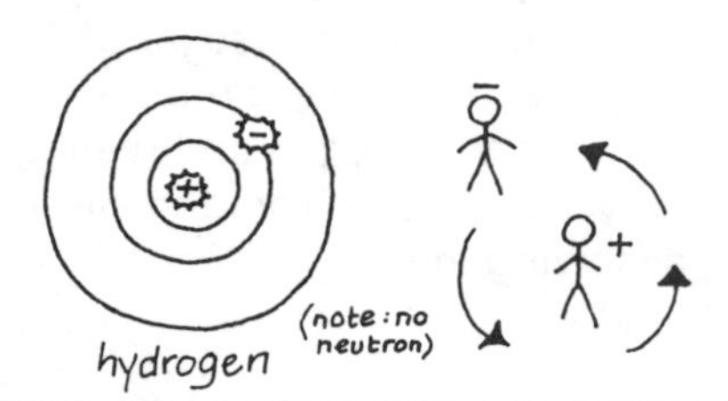

All atoms contain a nucleus (protons and neutrons) and electrons. Draw circles on desks or floors to represent the electron shells. Alternatively use students to represent the electron shells.

Models of molecules

Bottletop models

You will need
- *bottletops*
- *matchsticks*

Mark the bottletops on the underside with a pen or paint. Matchsticks form the bonds. Try to make all the examples in your textbook.

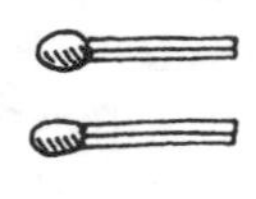

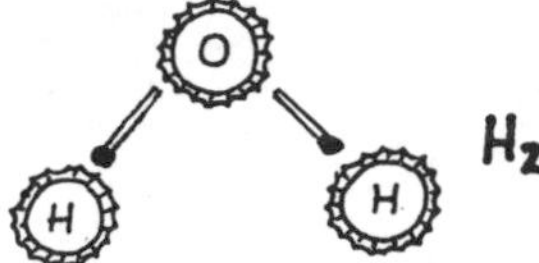

3-D models

You will need
- *small round objects, e.g. fruits, seeds, Plasticine, cork, foam pieces*
- *wire, string, sticks or matchsticks*

Use the wire, string etc. for the bonds and the fruits etc. for the atoms. More sophisticated 3-D models can be made using foam for atoms and matchsticks for bonds. Colour-coding the atoms of elements is also helpful.

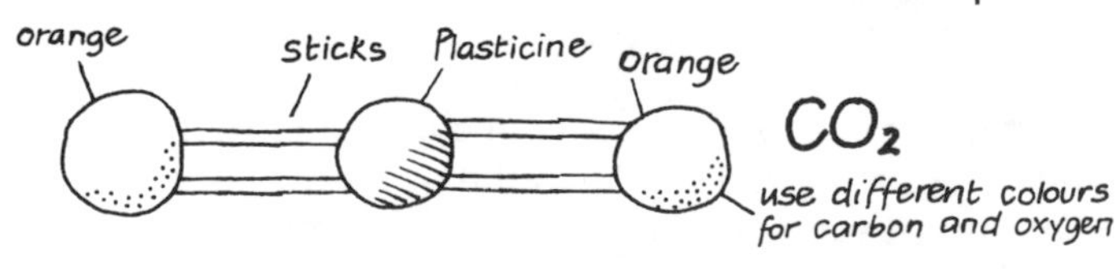

Common elements
recommended colour code for models of atoms and molecules

element	colour
carbon	black
chlorine	green
iodine	purple
hydrogen	white
nitrogen	blue
phosphorus	cream
sulphur	yellow
copper	gold
other metals	silver

Models of crystals

Crystal lattices

You will need
- *fruits, Plasticine, clay*
- *wire and sticks*

Make 3-dimensional models of crystals to show how the particles fit together. Look at textbooks for the crystal structures.

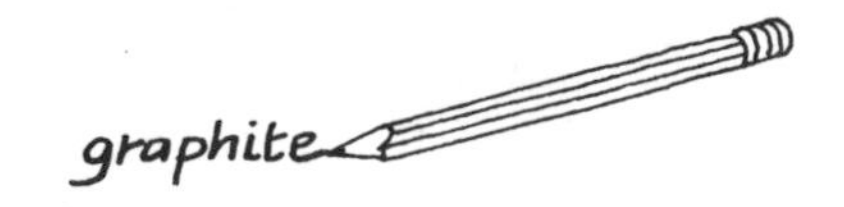

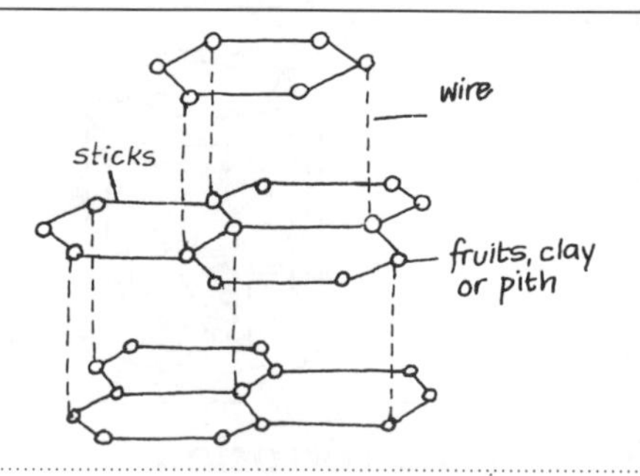

Lattice packing structures

You will need
- *small balls (must all be the same size), e.g. beads or peas*
- *clay*

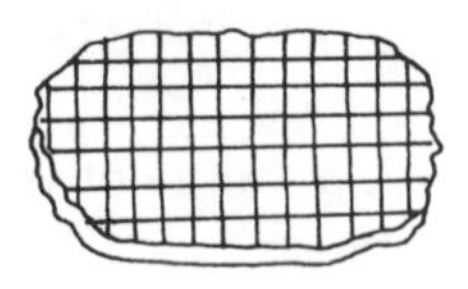

Lattice packing structures can be illustrated by allowing balls to settle in a container. They settle in an ordered pattern and are as close together as possible. Mark clay in a lattice pattern by scoring equal-sized pieces.

Bubble raft model

You will need
- *soapy water*
- *fine plastic tube*
- *shallow dish*

***Safety:* Do not allow students to share tubing because of the danger of cross-infection.**

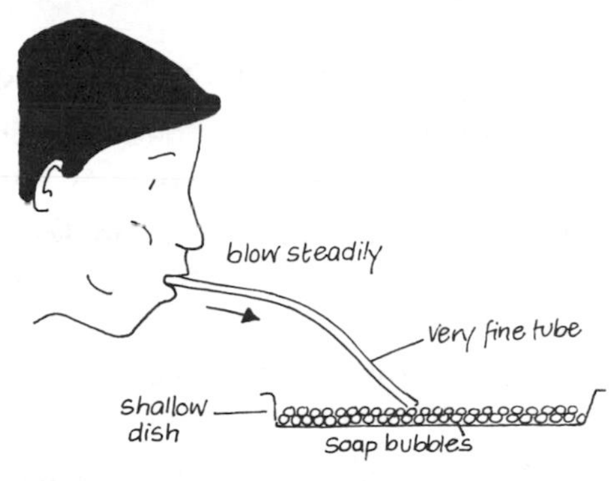

Put the soapy water into a shallow dish and blow. Tiny bubbles will form and pack tightly together in a lattice. The bubbles represent particles. The atoms of metals are arranged in this pattern.

Extend the activity by asking students what the 'faults' in the pattern could represent.

Making compounds from elements

Making sulphur dioxide

You will need
- *combustion spoon*
- *sulphur*
- *heat*
- *safety goggles*

***Safety:* Sulphur dioxide is poisonous.**

Heat the sulphur on the spoon and observe that it burns,with a blue flame and simply 'disappears'. The element sulphur has combined with oxygen in the air to make the compound sulphur dioxide. It has a characteristic smell.

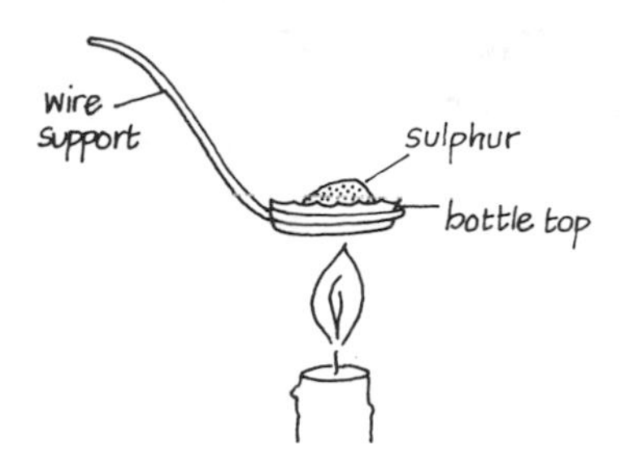

Making aluminium iodide

You will need
- *powdered aluminium*
- *powdered iodine*
- *bottletop*
- *safety goggles*

***Safety:* Poisonous iodine vapour may be given off, so perform the experiment outdoors or in a fume cupboard.**

Add 2–3 drops of water to make aluminium iodide which is grey-black.

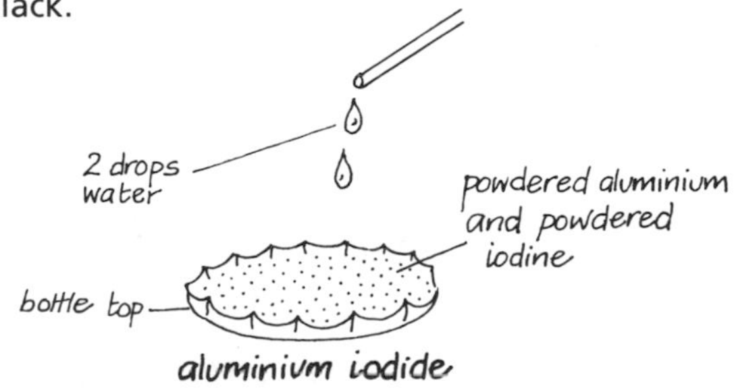

Making hydrogen

You will need
- *a metal*
- *a dilute acid*
- *test tube*
- *bottle with bung*
- *safety goggles*

Only use small quantities of metal and acid. Do not perform the experiment near an open flame. Always point test-tubes away from people. Do not use metals that are more reactive than magnesium.

Set up the apparatus as shown, adding the acid to the metal. (Magnesium and dilute hydrochloric acid work well.) Hydrogen fills up the test tube, displacing the air in the tube downwards. Demonstrate the gas is hydrogen by lighting it.

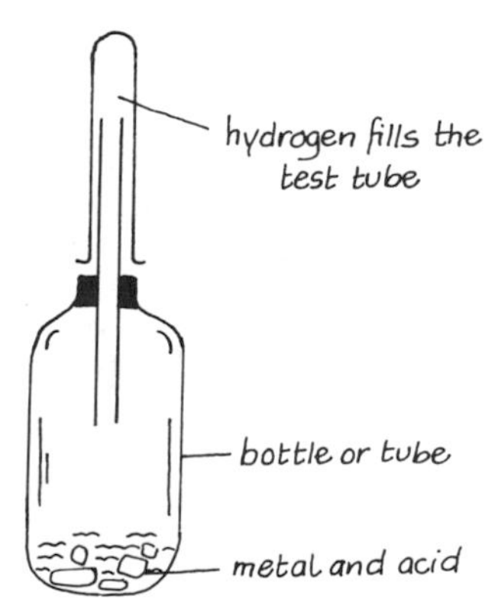

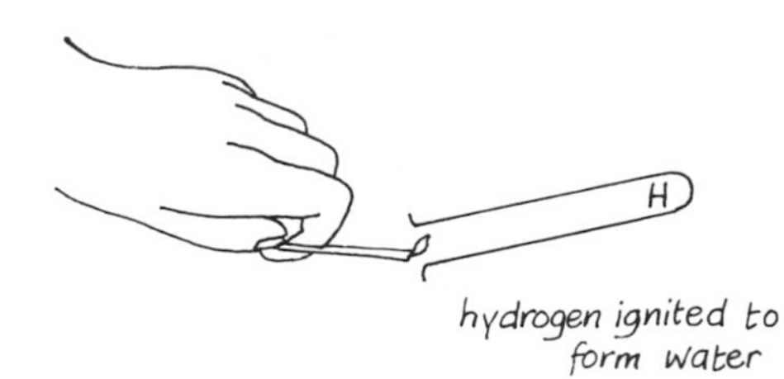

States of matter

- The **3 states of matter** are: **solid, liquid** and **gas**.
- Substances can be changed from one state to another by heating or cooling, e.g. ice → water → steam.
- **Molecules** are always moving. Molecules in solids only vibrate, in liquids they move freely, in gas they move around fast.
- If molecules are heated they move around more. A solid **melts** because the molecules vibrate so much that in the end they break away from each other and start to move freely – the solid melts.
- If a gas is cooled the gas molecules slow down and eventually the gas becomes a liquid – it **condenses**.

Introducing states of matter

Look for analogies in everyday life, e.g. bus stop, school assembly, places of worship.

A model for states of matter

You will need
- *a bottle*
- *seeds or stones*

A bottle with seeds or stones will act as a model.

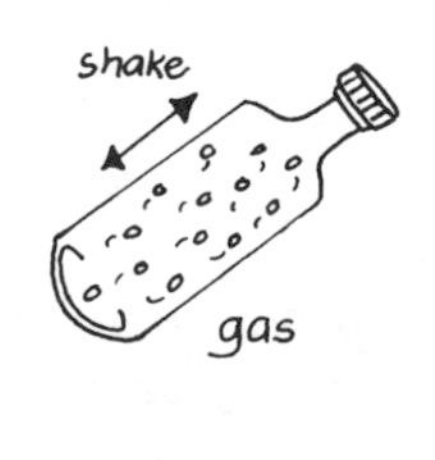

Solid to liquid

Comparing melting points

You will need
- *wax*
- *sulphur*
- *salt*
- *3 bottletops*
- *safety goggles*

Compare the time (and so the amount of heat) needed to melt equal amounts of wax, sulphur and salt.

Try other substances.

Some substances, e.g. waxes, are highly flammable when melted as they vaporise at low temperatures.

Determining melting point

You will need
- *tin can*
- *thermometer*
- *stirrer*

It would be safer to use a water bath in order to prevent the wax getting too hot and catching fire.

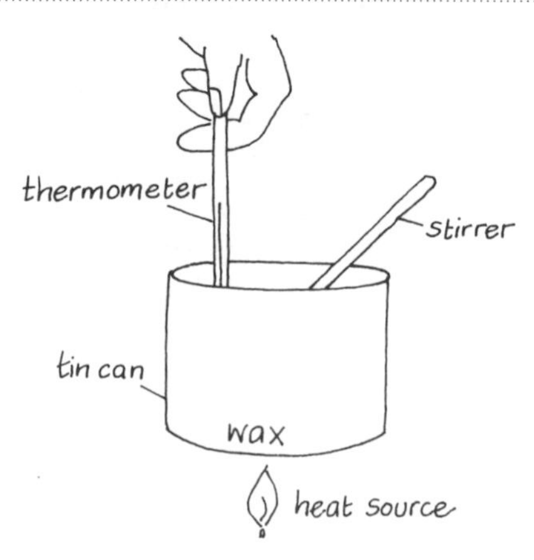

Gently melt the wax. Stir it continuously and ensure the thermometer does not rest on the bottom of the tin. Record the temperature at which all the wax has melted. You should notice that the temperature remains constant until all the wax has melted. The temperature will then rise. The point at which the temperature changes is the melting point.

Influence of surface area on melting

You will need
- *blocks of ice*
- *container*

Put some big blocks of ice into the container and record how long it takes before the ice melts. Now crush some more ice into smaller pieces. Compare the time it takes each to melt.

This could be done quantitatively if a thermometer is used.

Solids to liquids in the home

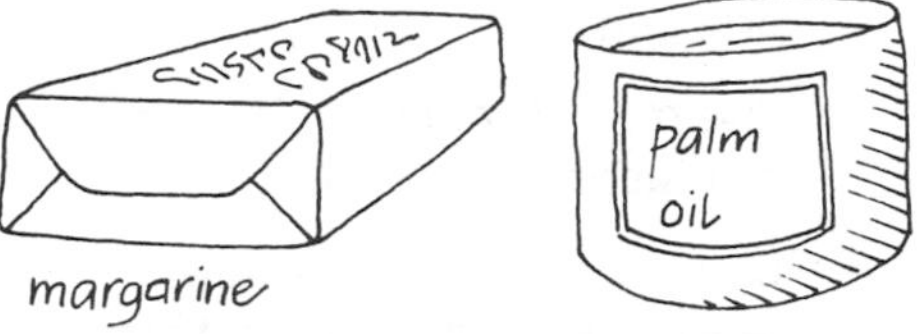
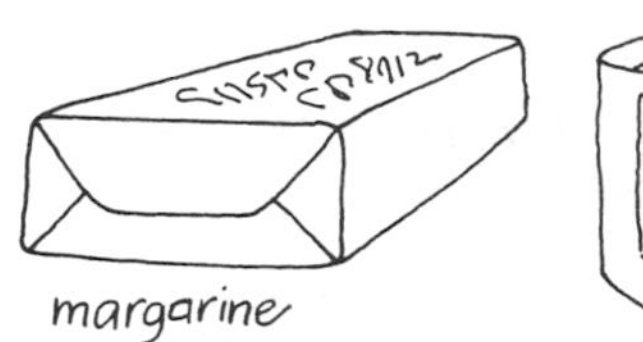

Liquid to gas

You will need
- *containers of different shapes*
- *water*

Compare the evaporation from containers with different surface areas and the same volume of water. The larger the surface area of water, the faster evaporation. An everyday application of this is in drying clothes.

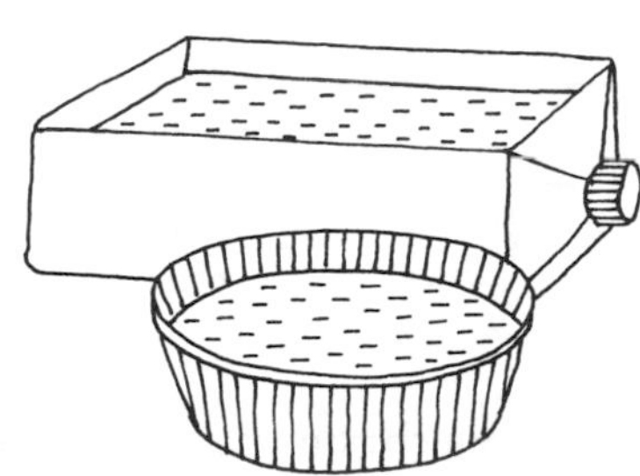
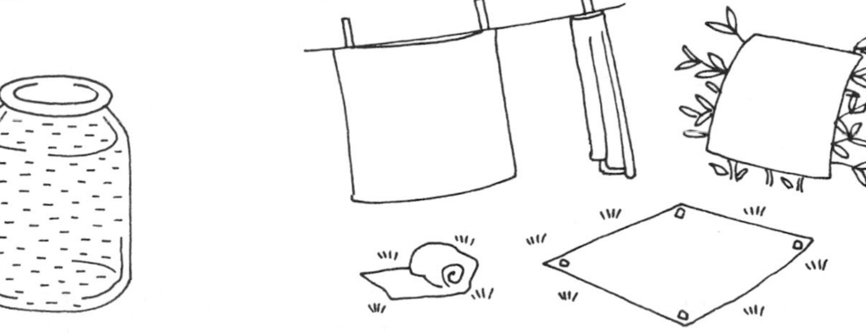

Gas to liquid

***Safety:* Do not allow students to share straws or ballpoint casings because of the danger of cross-infection.**

There are many situations where condensation can be seen.

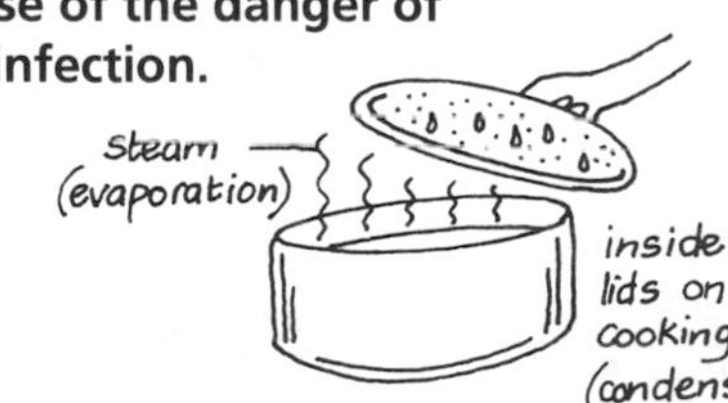

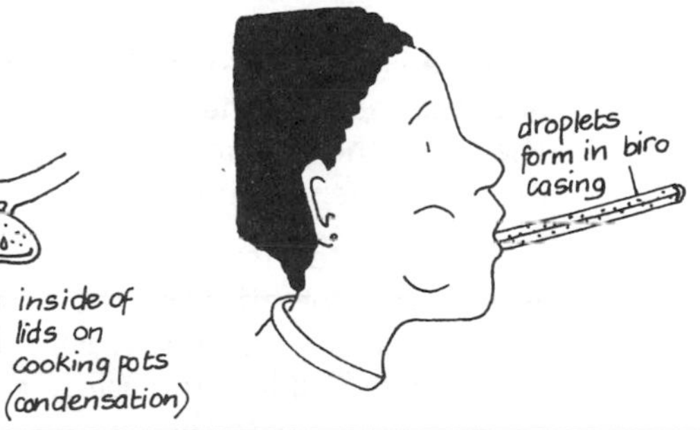

Liquid – gas – liquid

Distillation

You will need
- *a large metal can*
- *safety bung (cork or rubber)*
- *cork*
- *plastic tubing*
- *wet cloth*
- *container*
- *source of heat*
- *water*

Ensure the can always contains water and that the safety bung is not too tight.

This is an application of changing gas to liquid by the process of condensation.

Heating the can will produce steam which is then cooled by the wet cloth. Steam condenses to produce water. This method could be used to purify water.

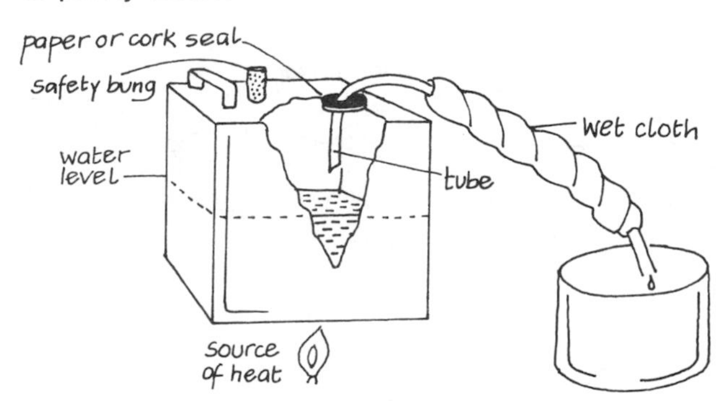

Speeding up reactions

The rate of a chemical reaction depends on many factors. Some examples are given below.

- Increasing the surface area of **reactants** (starting materials) increases the **rate of reaction**.
- The higher the **temperature**, the faster a **reaction** proceeds.
- The higher the **concentration** of reactants, the faster the reaction proceeds.
- The **rate of reaction** slows down as the concentration of the **product**(s) increases.
- **Catalysts** or **enzymes** (biological catalysts) speed up reactions. Some reactions proceed so slowly without a catalyst that they do not seem to take place at all.

Sugar and water

Investigating surface area

You will need
- *card*
- *scissors*
- *sugar cubes*

Students can find the surface area of the cube by drawing around the cube's faces.

Investigate with students the relationship between volume and surface area by doubling and tripling the volume.

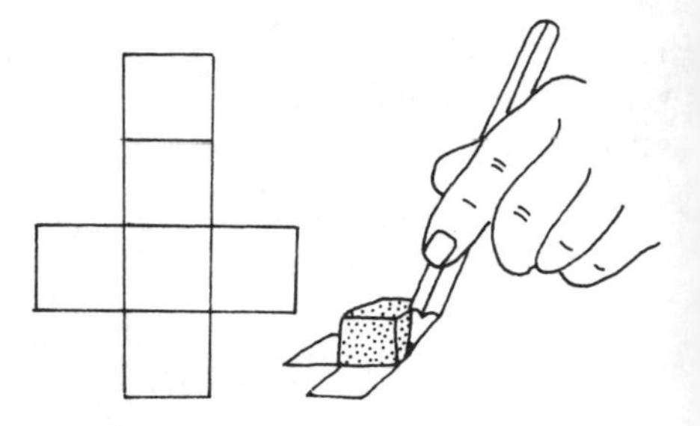

Sugar and water

You will need
- *sugar cubes*
- *water*

Drop a sugar cube into one container. Drop a crushed sugar cube into another. Note which dissolves fastest. Ask students why this is so.

Extend the activity by investigating the effect of raising or lowering the temperature on the speed of dissolving. Link this to cooking.

Limestone and acid

You will need
- *limestone or crushed shells*
- *acid*
- *thermometer*
- *safety goggles*

Add the limestone to the acid and note the speed of the reaction. The faster the reaction the faster the limestone will fizz.

Alter one factor after another and observe the effect on the rate of reaction. For example: crush the limestone; dilute the acid; increase or decrease the temperature of the acid.

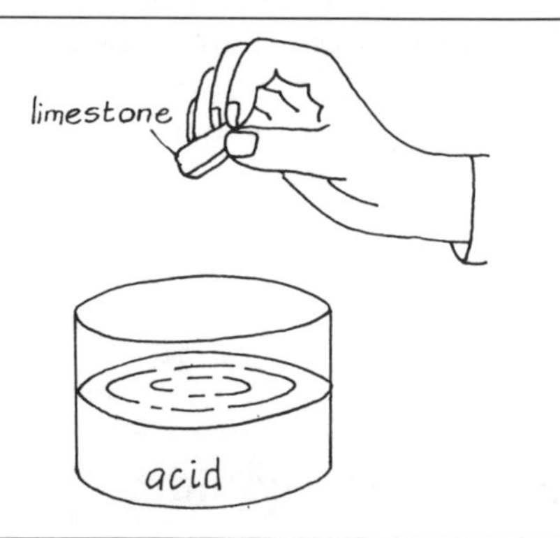

Measuring rate of gas evolution

You will need
- *flexible bottle*
- *plastic tube*
- *graduated measuring cylinder or bottle*
- *safety goggles*

The apparatus can be used to remove known quantities of gas from the collecting bottle. Alternatively if gas is simply collected in a large enough calibrated bottle there is no need to remove gas.

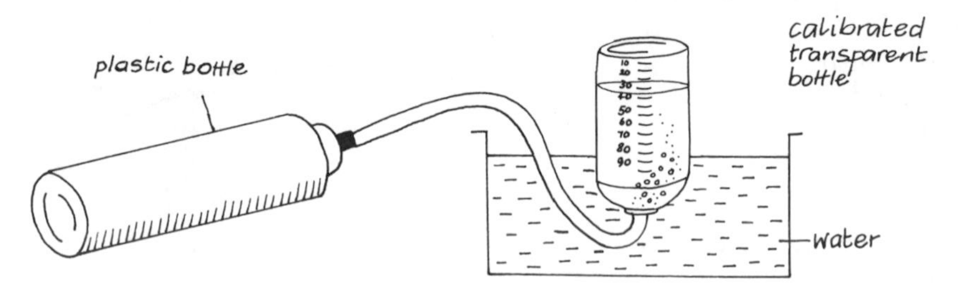

Baking powder and acid

You will need

- *baking powder*
- *vinegar or dilute acid*
- *safety goggles*

Add baking powder to the acid. The mixture will fizz as carbon dioxide is produced. The more violent the fizzing, the faster the reaction. Alter one factor at a time and discuss why the rate of reaction changes. For example, alter the concentration of reactants and the temperature.

This experiment could be quantitative if the gas were collected and measured.

Hydrogen peroxide

With an enzyme

You will need

- *hydrogen peroxide*
- *soap or detergent*
- *large container*
- *enzyme*
- *safety goggles*

Enzymes from crushed potato, liver or yeast are suitable. Place a small piece of raw potato or liver in the mixture. If the potato is hard, it may help to crush it first. If using dried yeast, activate it (follow instructions on the packet).

As oxygen is produced, foam is produced on the surface. The height of the foam can be used as a measure of the speed of reaction. Investigate changes in the speed of reaction caused by adding more of the raw potato or liver. This increases the amount of the active enzyme peroxidase. Alternatively, try adding more soap.

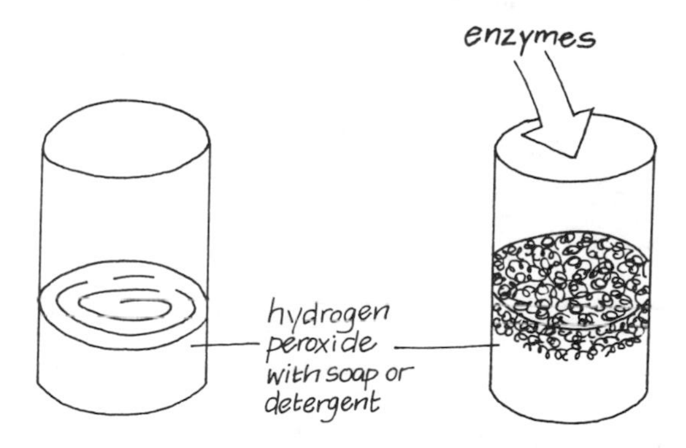

With a chemical catalyst

You will need

- *manganese(IV) oxide*
- *hydrogen peroxide*
- *injection bottle*
- *collecting bottle and tube*
- *safety goggles*

Manganese(IV) oxide is the black paste in dry cells.

Mix the reactants and collect the gas (oxygen) over water. This reaction is suitable for investigating

- concentration
- effect of a catalyst

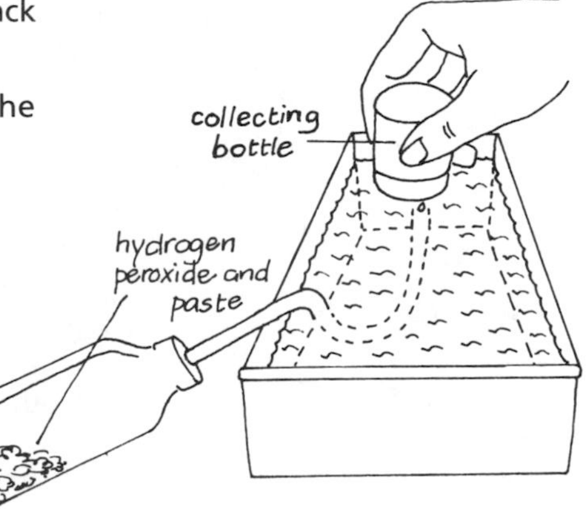

Acids and bases

- Acidity and alkalinity is measured in **pH**. A pH of 7 is **neutral**, less than 7 is **acid**, greater than 7 **alkali**.
- An alkali is a solution of a **base** in water. Not all bases are **soluble** in water.
- **Indicators** are chemicals whose colour depends on whether they are in an acid or alkali **solution**. They can be used to measure pH.
- Acids react with **metals**, **bases** and **carbonates** to form **salts**.
- Acids are **neutralised** when they are 'used up' in a reaction, e.g. an alkali can neutralise an acid, but if too much is added the solution will then be alkaline not neutral.

Making indicators

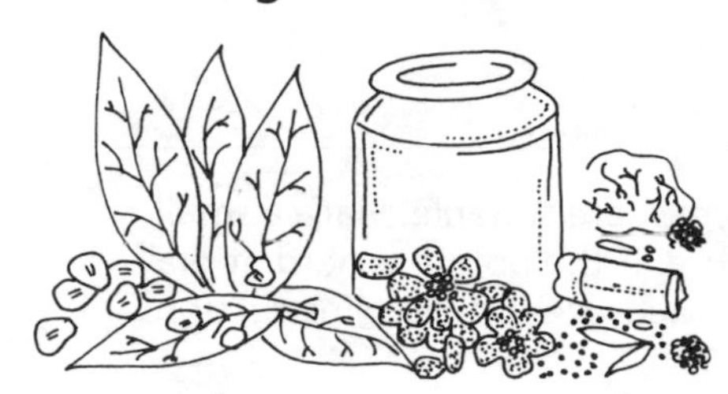

Many red, violet, yellow or pink flowers or fruits and leaves can be used as indicators. They should be crushed with water or colourless spirit. Spirit-based indicators are more stable. Boiling improves the extraction of colour.

Students could investigate which local flowers, leaves etc. produce the most effective indicators.

The accuracy of home-made indicators could be compared with commercially produced indicators such as litmus or methyl orange.

Every-day acids and bases

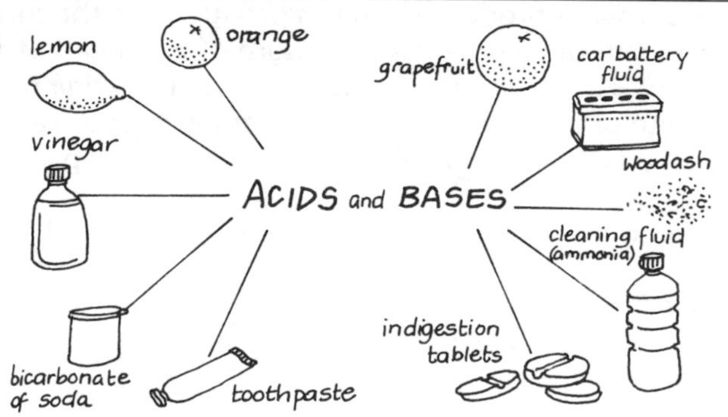

Many substances in everyday life are either acids or alkalis. Some examples are shown here.

Ask students to test some of these items so they can identify the acids and alkalis using indicators.

Acid rain

Pollution e.g. from power stations, factories and cars is carried in the wind. The acid rain may fall a long way from the cause of the pollution – often in a different country.

Making salts

Acid and carbonate

You will need
- *hydrochloric acid*
- *limestone*
- *baking powder*
- *egg shells*
- *safety goggles*

Mix the hydrochloric acid with each of the other substances, which are all carbonates. In each case the mixture will bubble and froth (showing a reaction is taking place) and then the liquid clears. The acid has combined with the carbonate to produce a salt, in this case sodium chloride or 'common, table salt'.

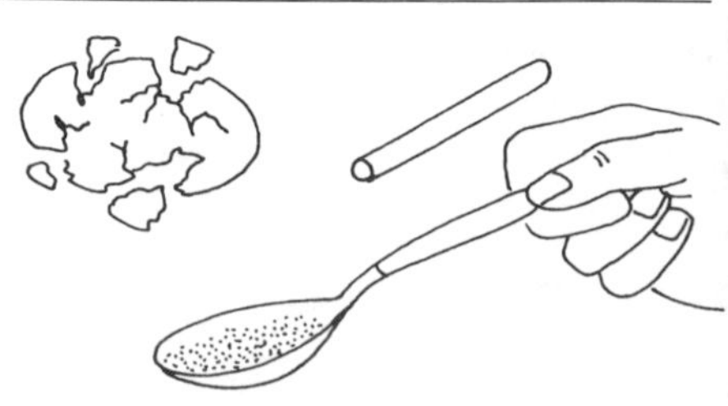

Acid and metal

You will need
- *dilute acid*
- *samples of metals*

Metals will react with the acid and appear to be 'eaten away'. The stronger the acid the more obvious the reaction. If dilute solutions are used the reaction may take several days to be obvious.

Changes in pH can be observed by putting an indicator into the acid.

Acid and alkali

You will need
- *solution of potash*
- *vinegar*
- *indicator*
- *transparent container*
- *safety goggles*

Add woodash to water, stir and then filter. The clear solution is a solution of potash (potassium carbonate). Add enough indicator for the colour to be obvious. Add the vinegar slowly until the colour of the indicator changes. The amount of acid required to cause a complete colour change is the quantity which neutralises the alkali. If more acid is added the solution will become acid.

The experiment can be done quantitatively, as a titration.

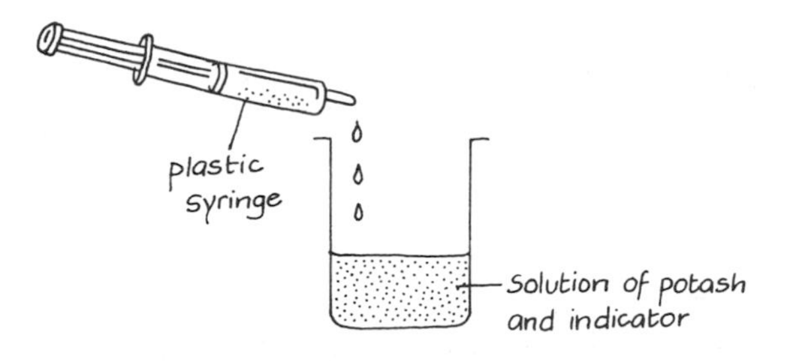

Bee stings

Bee stings are acid. To soothe the pain, a weak alkaline solution is used to neutralise the acid of the sting.

Growing salt crystals

You will need
- *Epsom salts (magnesium sulphate)*
- *magnifying glass*
- *glass dish*

Dissolve some Epsom salts in water. Pour a few drops into a shallow glass dish and put it into sunlight to evaporate. Examine the crystals which form with a magnifying glass.

A salt garden

You will need
- *vinegar*
- *salt*
- *coal, porous rocks or broken pottery*

Make a saturated salt solution. To do this dissolve salt (common table or cooking salt) in water until no more will dissolve. The solution is then a saturated solution. Gently add a spoonful of vinegar to the surface of the solution. After a few days delicate crystals form around the rough surfaces of the rocks or pottery. Be careful not to disturb the liquid or the effect will not be as dramatic.

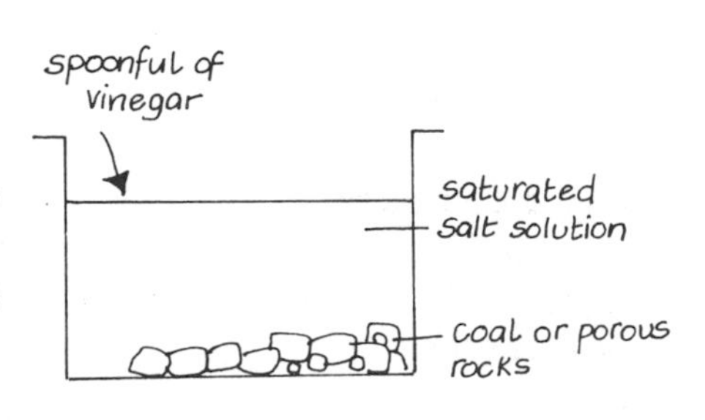

Magnetism

- **Magnetism** is the **force** by which iron, cobalt and nickel attract other metals. Unless they are **magnetised** metals cannot attract other metals.
- Magnets have **north** and **south poles** (ends). If a magnet is suspended freely it will align with the **Earth's magnetic field,** i.e. with the magnetic North and South Poles.
- If 2 magnets are brought close together, the **like poles** will **repel** each other, **unlike poles** will **attract**. (See static charge page 76.)
- Magnets can be created either by using another magnet or by using electricity.

Magnetising objects

Making a big magnet

You will need

- *an iron bar*
- *hammer*
- *two strong magnets*

Hit the iron bar several times on one end between two strong magnets lined up north–south. This will magnetise the bar. Test it by picking up small nails. (Try pointing the bar in a north–south direction while hitting it.)

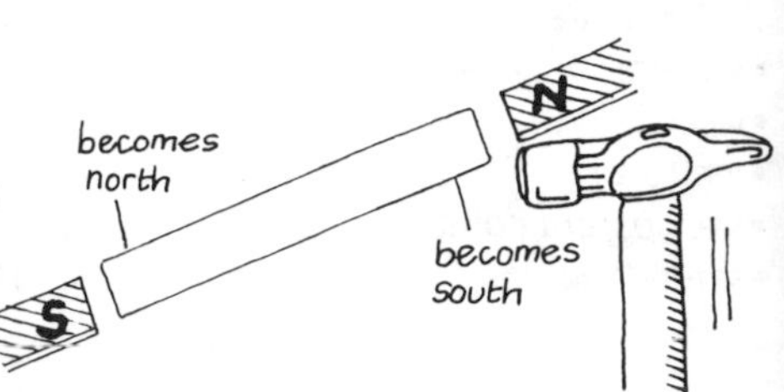

Magnetising iron or steel

You will need

- *a bar-shaped magnet*
- *nail (iron or steel)*

Stroke the nail from end to end with one end of the magnet.

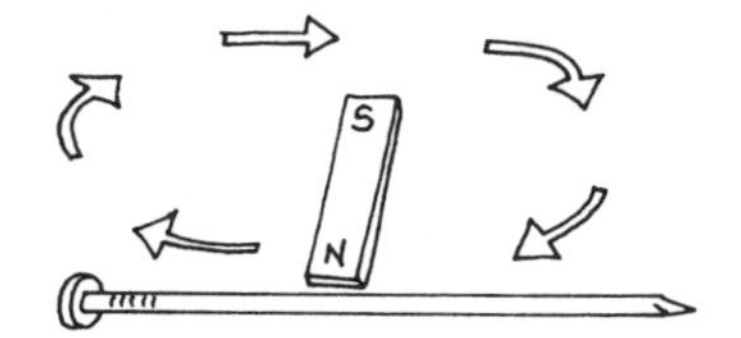

Magnetism from electricity

You will need

- *dry cell*
- *insulated copper wire*
- *iron nail*
- *small nails or pins*

Make about 50 turns of wire around the nail. Connect the wire to the dry cell. Pick up pins with the magnetised nail.

Vary the number of turns of wire and investigate the effect on the strength of the magnet.

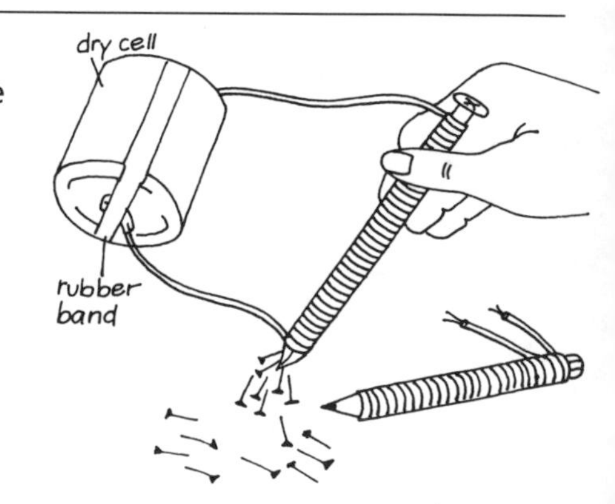

Investigating lines of force

Using nails

You will need

- *bar magnet*
- *glass*
- *small nails or pins*

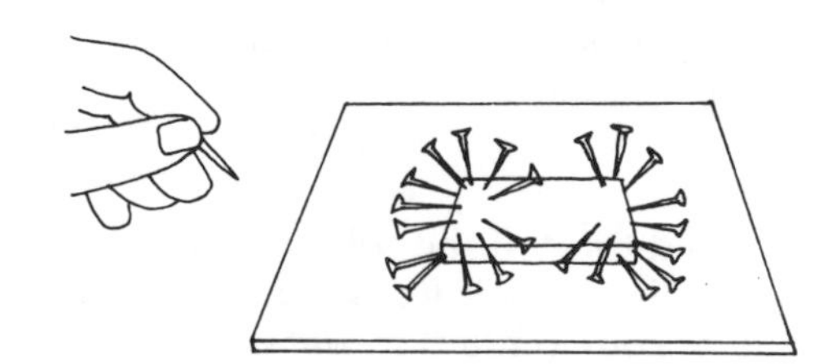

Place the glass over the magnet and let the nails fall onto it one at a time. The nails make a pattern following the lines of force.

Using a compass

You will need

- *a compass*
- *insulated copper wire*
- *dry cell*

Connect the coiled wire to the cell. Move the compass and record changes in the direction of the pointer. The pointer aligns along the lines of magnetic force and these can be followed or traced.

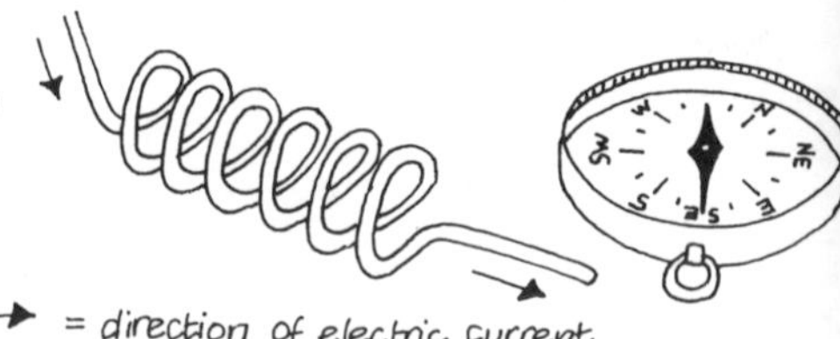

Magnetism through materials

You will need
- *glass container*
- *paper*
- *nails or pins*
- *water*

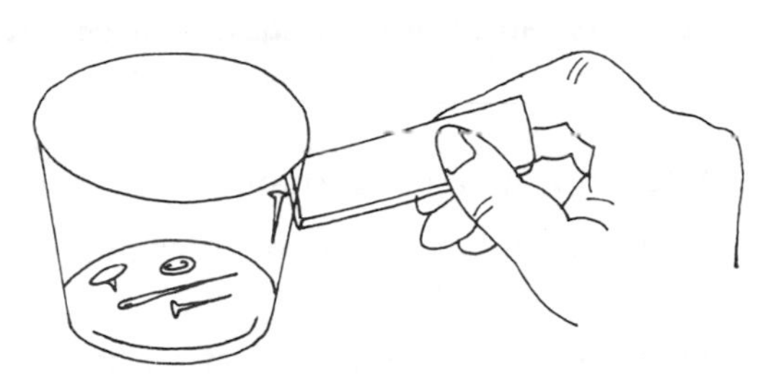

Put the nails inside the container and test if they are still attracted to a magnet held outside the glass. Put water in the container and test again.

As an extension, test different materials to see which allow magnetism through. Also test materials of different thickness.

Looking at magnetic fields

You will need
- *thin card*
- *iron filings*
- *large magnet*
- *spray apparatus shown*

Safety: **Do not allow students to share straws because of the danger of cross-infection.**

Put the magnet under the card and sprinkle iron filings on top of the card. Tap the card gently if the pattern is not clear. Once the pattern has settled, spray salt water over the iron filings. Repeat after several hours and leave overnight. A rust print will develop.

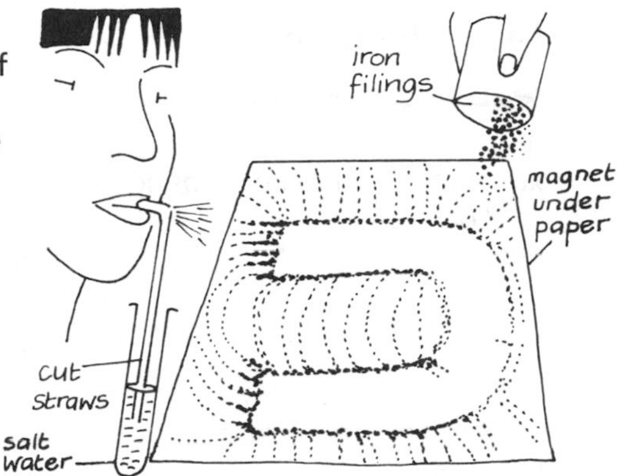

Earth's magnetic field

You will need
- *a fine bar magnet*
- *thin thread*
- *support (non-metallic)*

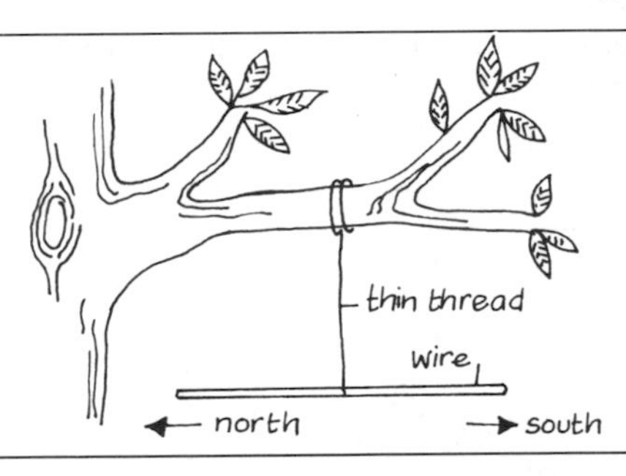

Use iron or steel wire to make the bar magnet. Old coathanger wire works well (unless it is aluminium!). Suspend the magnet and leave it overnight. The wire will point north–south in the morning. Beware of air movements.

Compasses

Water compass

You will need
- *non-metallic dish of water*
- *magnetised needle*
- *paper or thin card*

Rest the needle on the card and float the card on the water. Leave the paper to float freely and it will eventually align north–south.

Try other types of raft, e.g. corks.

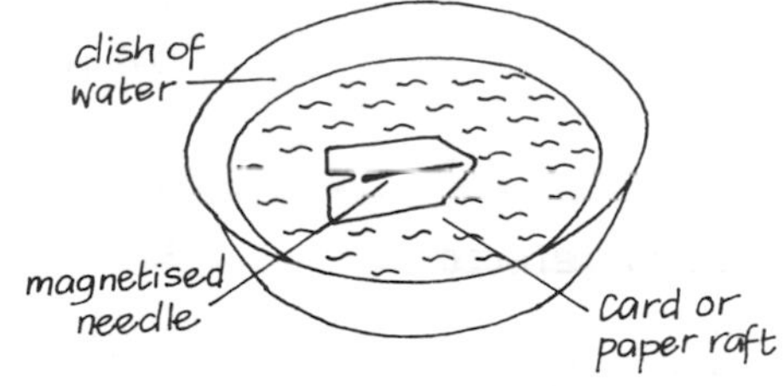

Button compass

You will need
- *magnetised needle*
- *round-faced button*

Place the button on a smooth surface and note that it will align north–south. The button must be able to move very freely for the force of magnetism to overcome that of friction.

Ask students to identify the north and south poles of the needle. Using a bar magnet ask them to test that like poles repel each other, opposites attract.

Suspended compass

You will need
- *branched stick*
- *fine thread*
- *magnetised needle*
- *sand-filled non-metallic container*

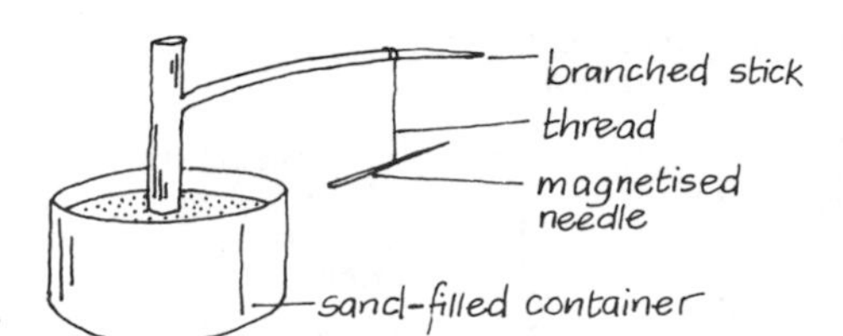

The needle will eventually align with the Earth's magnetic poles, i.e. north–south.

Electricity

- **Static electricity** is a **charge** which can be created on a surface. Static electricity can be either **negative** or **positive** depending on the materials used to create the charge.
- Static electricity **attracts** oppositely-charged materials and **repels** materials with a similar charge (see magnetism page 74).
- **Electricity** is a flow of charge. A **circuit** has to be complete for electricity to flow.
- Electricity can be made by **electrolysis**.

Static electricity

Creating a charge

You will need

- *plastic comb or ball point casing*
- *wool or nylon*
- *paper*

Add charge to the comb or ballpoint casing by rubbing on wool or nylon. Tear the paper into small pieces. Hold the comb or ballpoint casing above the paper. The paper is attracted to the 'charged' plastic and sticks to it.

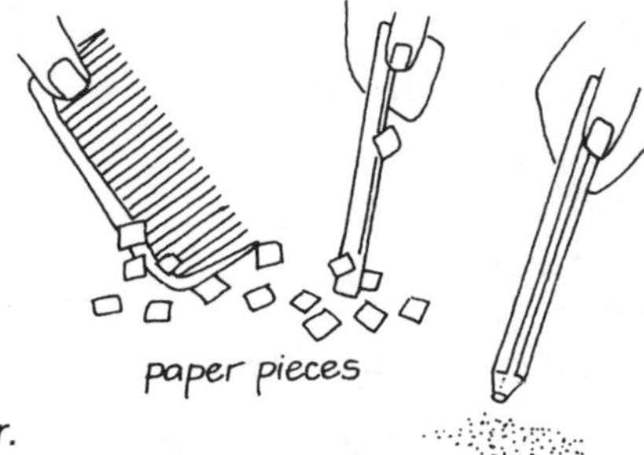

Try salt and pepper instead of paper.

Observing the effects of charge

You will need

- *fine thread (ideally silk)*
- *pith or polystyrene ball, ping pong ball or puffed rice grain*
- *wooden support*
- *charged comb*

Suspend the ball(s) by the fine thread from a wooden stick. Hold a charged comb near the ball and notice that the ball is either repelled or attracted by the comb. A pith ball can be made by using pith from the inside of a suitable plant stem. Dry the pith and press into balls. (Balls of 5 mm diameter work well.)

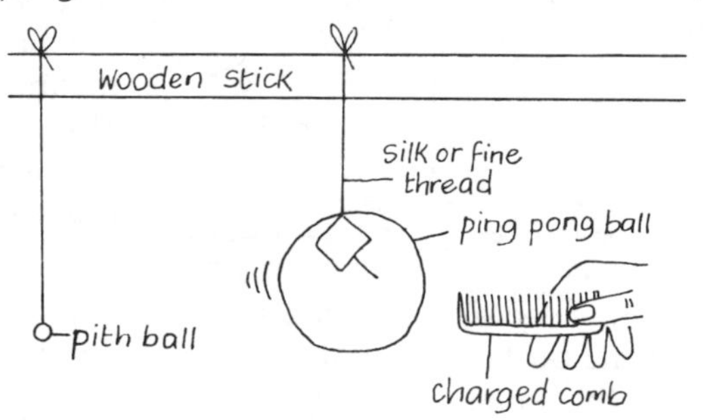

As an extension, suspend 2 inflated, charged balloons near to each other on the stick. Also charge a comb by rubbing on wool and then hold it near water running slowly from a tap or a hole in a can. In both cases, ask students to note what they see and explain why it happens in terms of charge.

Simple detector

You will need

- *sand-filled can*
- *mounted needle*
- *paper strip*
- *charged comb*

Mount the needle as shown and balance the paper strip. When a charged object is held near to the paper the paper moves. It is affected by the static electricity on the comb.

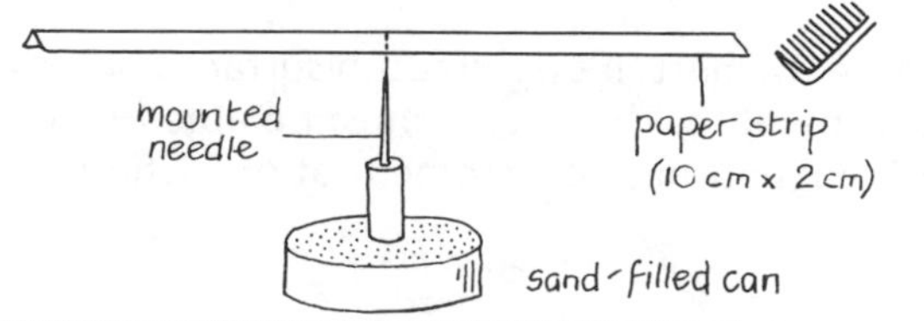

Ask students to note which direction the paper moves. Ask whether the movement indicates the object has the same or opposite charge to the paper.

Light bulb electroscope

You will need

- *light bulb*
- *wire*
- *fine foil, e.g. cigarette foil*
- *thin hard plastic sheet*
- *support, e.g. jar lid*

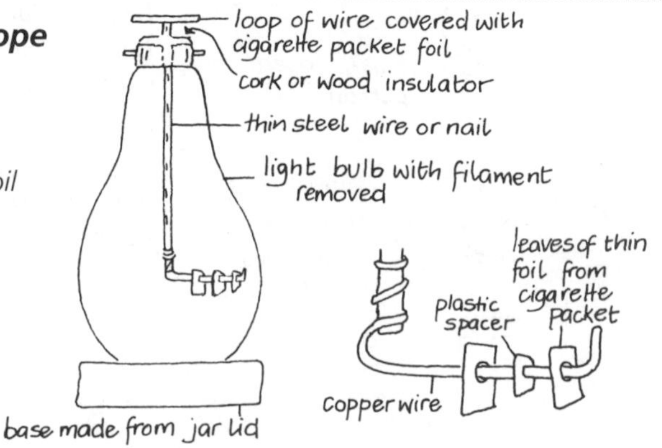

Make the apparatus as shown. When a charged object is brought close to the loop of wire at the top of the light bulb the foil will move. You may be able to see whether the foil moves away from, or towards the object.

Making electricity

Simple cells

You will need

- *a jar*
- *dilute acid, e.g. sulphuric acid*
- *2 metals, e.g. copper and zinc*
- *voltmeter*
- *safety goggles*

Set up the apparatus as shown.

This experiment is useful to investigate the reactivity series (see page 64). The further the metals are apart in the reactivity series the bigger the voltage.

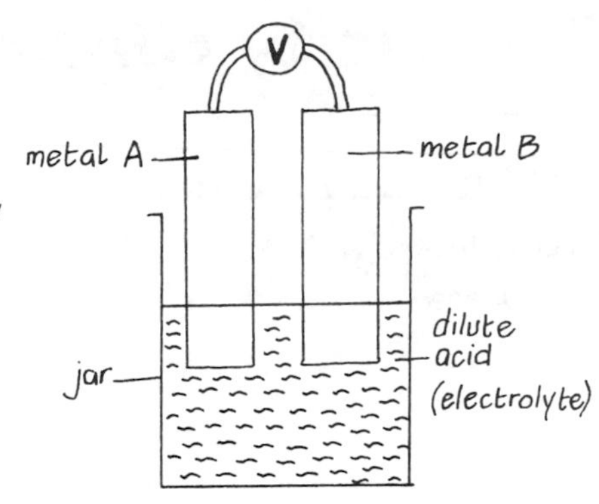

Electricity from fruit

You will need

- *citrus fruit, e.g. lemon*
- *zinc strip*
- *copper strip*
- *copper wire*
- *torch light bulb*

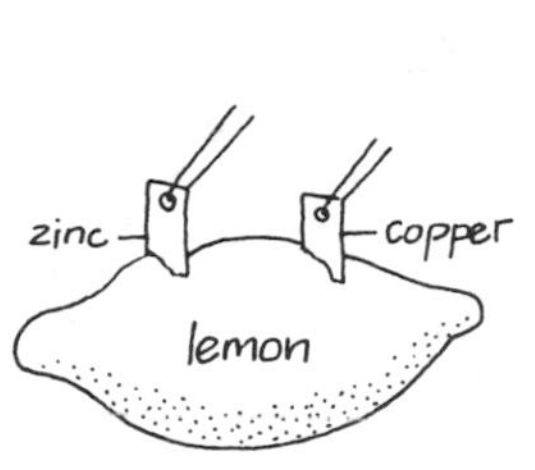

or

Make the circuit shown. Make sure the electrodes (metal strips, drawing pins or paper clips) do not touch. Any citrus fruit has acidic juice and so could be used in this experiment.

Ask students to see if using several fruits in series produces more electricity – indicated by the bulb shining more brightly.

Inside cells and batteries

A cell has 2 electrodes only. A battery is, strictly speaking, a group of cells. Confusingly, what is often called a battery in a shop is, in fact, a cell!

Inside a cell

You will need

- *an old dry cell (often called a battery)*
- *safety goggles*

Remove the outer coating and cut the inside in half so the components can be seen clearly. Many of the contents can be useful.

Ask students to identify the electrolyte, the cathode and the anode.

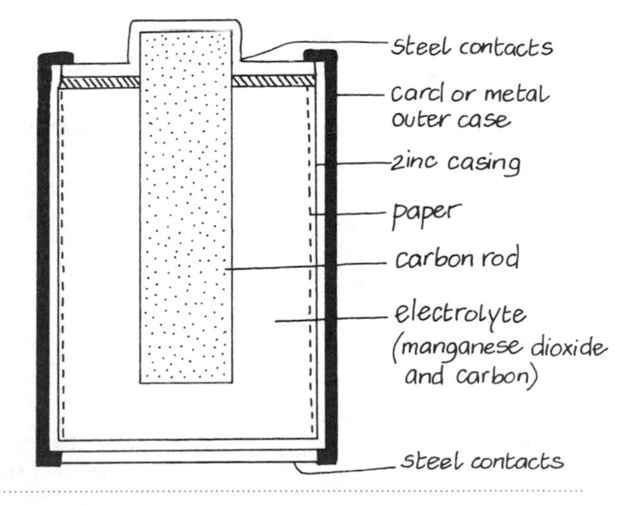

***Safety first:* Care must be taken while cutting through the cell.**

Car batteries

Acid from car batteries is very strong and will cause skin burns.

Car batteries contain lead plates as the electrodes and sulphuric acid as the electrolyte. The electrodes are in pairs.

Investigate why electricity is produced, since both electrodes in a pair are made from lead.

Electricity continued

Making equipment

Circuit boards

You will need
- *nails*
- *board*

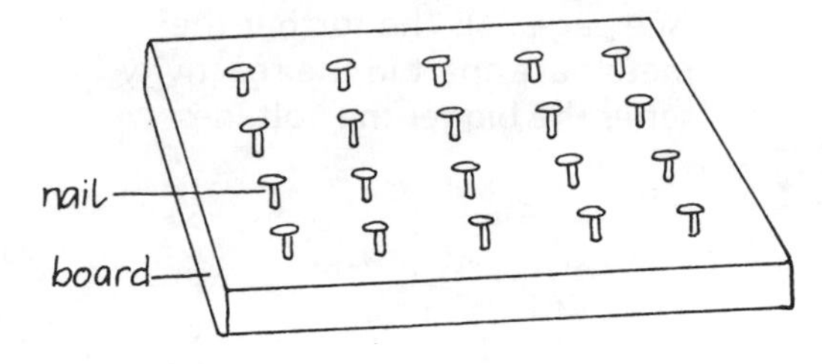

Light bulb holders

You will need
- *heavy wire*
- *bent strips of metal, e.g. from a can*
- *bottletops*

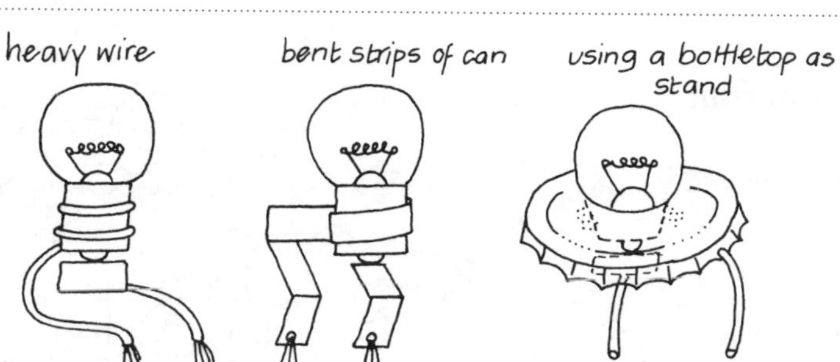

Switches

You will need
- *nails or drawing pins*
- *small boards*
- *wire*
- *paperclips*

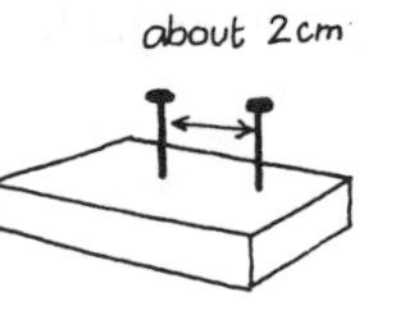

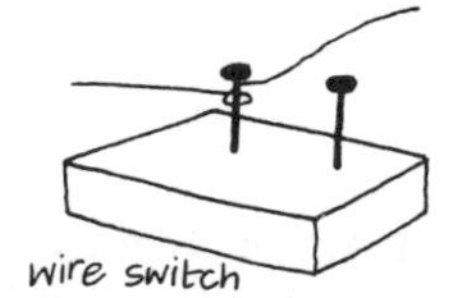

Three nails and a clip will make a two-way switch.

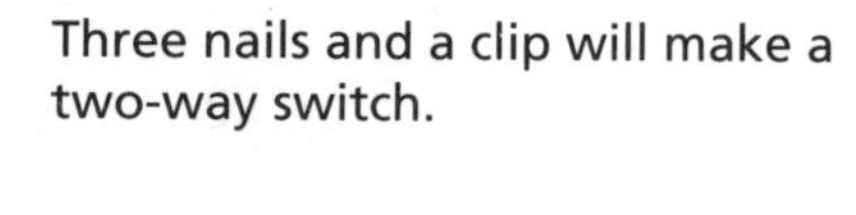

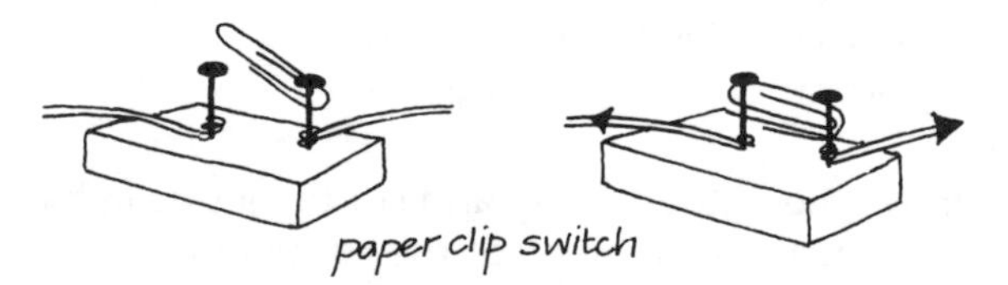

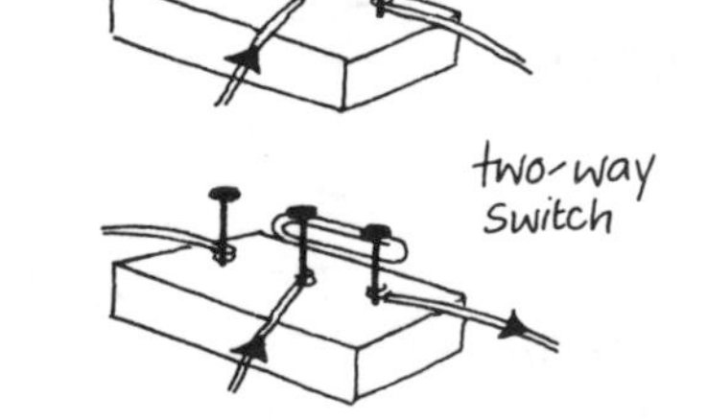

Rheostat

You will need
- *dry cell*
- *metal strip*
- *pencil*
- *wire*
- *torch light bulb*

When the lever is moved to the left along the graphite of the pencil 'lead' the torch bulb burns more brightly.

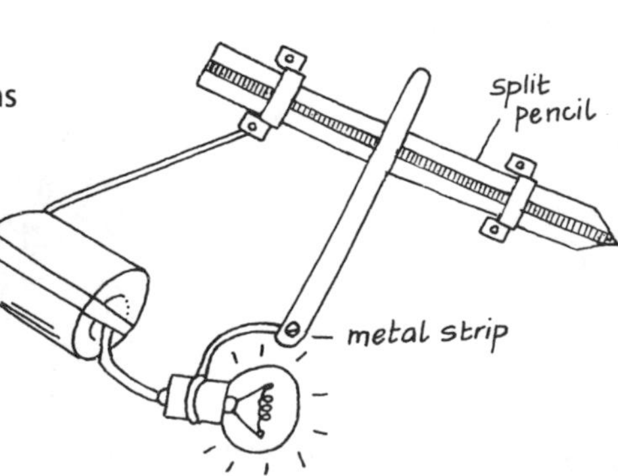

Wires

Find wires in old radios or other equipment. Use aluminium foil wrapped over thin wire to make it thicker.

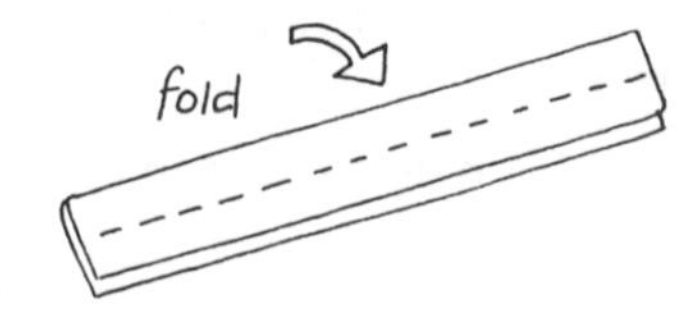

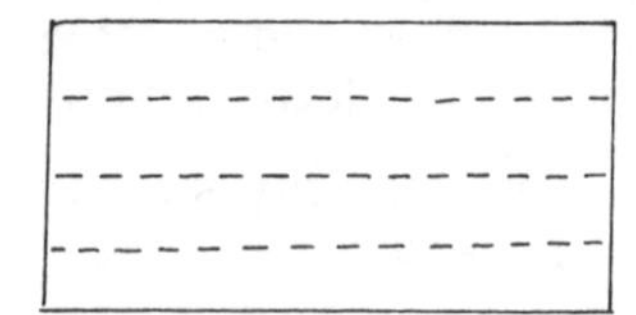

Useful analogies

Switches

The drawbridge acts as a switch.

two-way switch

The plank can only be in one of two positions. It is analogous to a two-way switch.

Circuits in series

If the bridge breaks, the flow stops, i.e. if one component breaks, the circuit is incomplete and electricity cannot flow.

Circuits in parallel

If one bridge breaks the race can go on, i.e. if one component fails there is still an alternative route for the electricity to flow.

faster pace

Water as an analogy for electricity

The river (electricity) flows through the narrow and the wide part of the river. However, where the river is narrow the amount of water flowing (the current) is smaller, but the resistance or power is greater, while the voltage stays the same.

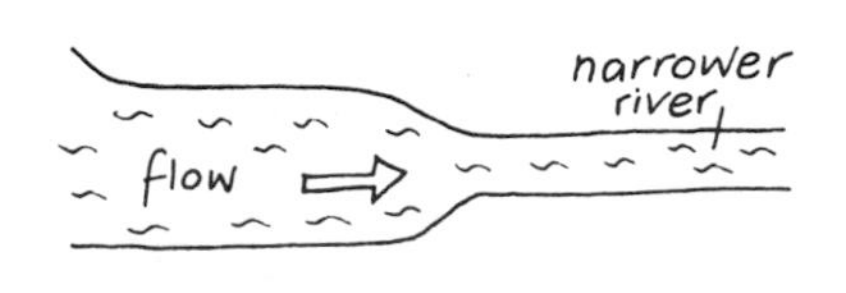

A dam acts like a switch. Unless the dam is opened no water can flow.

Investigating a torch

You will need
- *torch or flashlight*

Take the torch apart. Identify the function of each of the parts, e.g. identify the conductors, insulators, switch, etc. Discuss how the torch works as it is reassembled.

Electric motors

- **Electricity** flowing through a wire creates a **magnetic field** around the wire.
- If electricity is passed through a coil of wire, the coil acts like a **magnet** with **north and south poles**.
- **Magnetic forces** of like charge (poles) repel each other, while opposite charges (poles) attract each other.
- Reversing the direction of the **electric current** reverses the direction of the magnetic field.

Compass and electromagnet

You will need
- *a compass*
- *electromagnet (see page 74)*

Place the electromagnet close to a compass needle. Turn the current to the electromagnet on and off. If the current flows with no break, the compass needle moves and then stays still. When the current is switched off, the needle returns to its previous position.

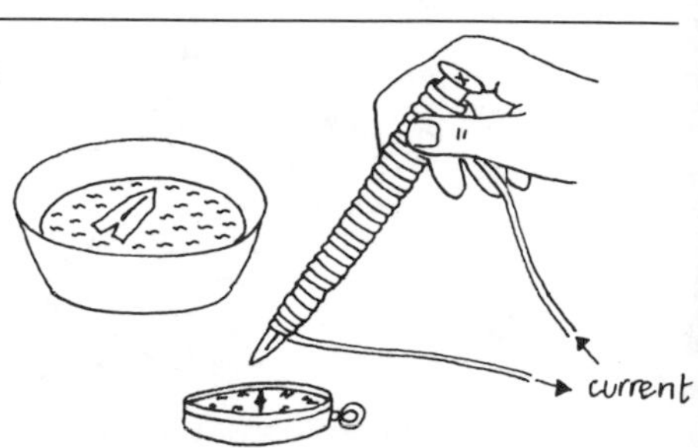

Magnet and movable conductor

You will need
- *copper wire*
- *bar magnet*
- *2 nails*
- *wooden base*
- *support wire*

Set up the apparatus as shown. When a current is passed through the copper wire, the wire moves. Make sure the copper wire is clean and that the ends in contact with the loops make a good connection. If contact or the reaction is poor then try altering the shape of the support loops.

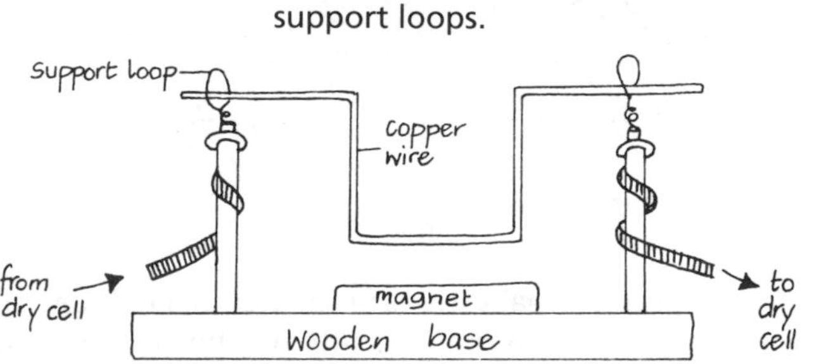

Experiment with the apparatus by reversing the magnet poles, then reverse the direction of the current. Also try using wire of different thickness.

Making electric motors

In a simple motor a coil is made to spin round. If the spinning coil is attached to wheels, for example, the wheels would turn when the coil turned.

Coil mountings

You will need
- *materials for support loops*
- *copper wire*

Coils can be supported in many ways. Some are shown here.

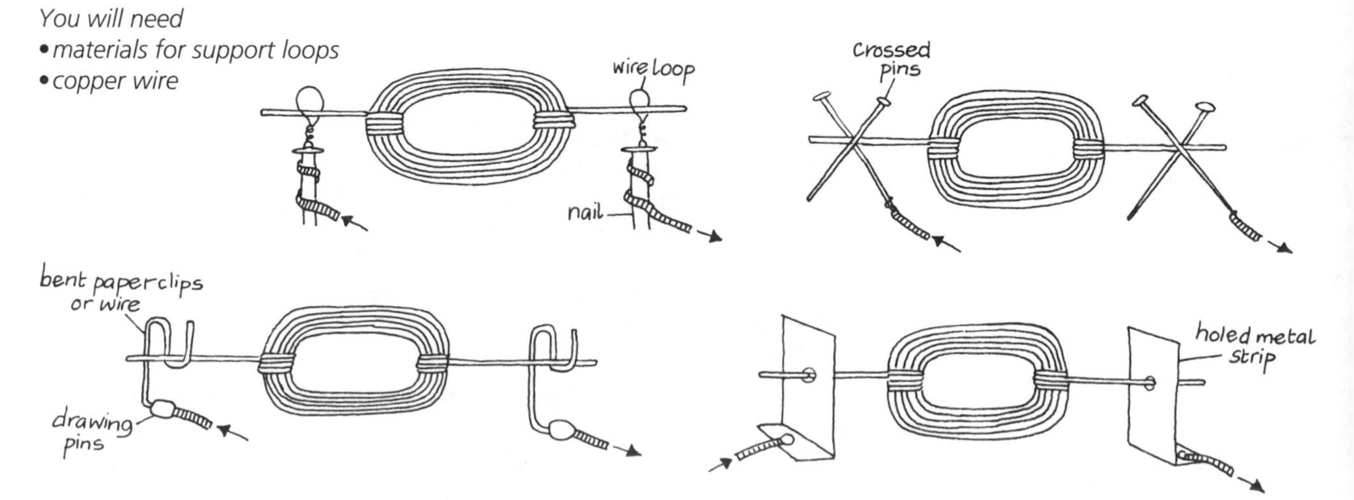

Magnet mountings

You will need
- *mounting materials*
- *2 bar magnets*

The magnets must be supported on either side of the coil. Two methods are shown here. A wooden base board could be made to mount the magnets permanently.

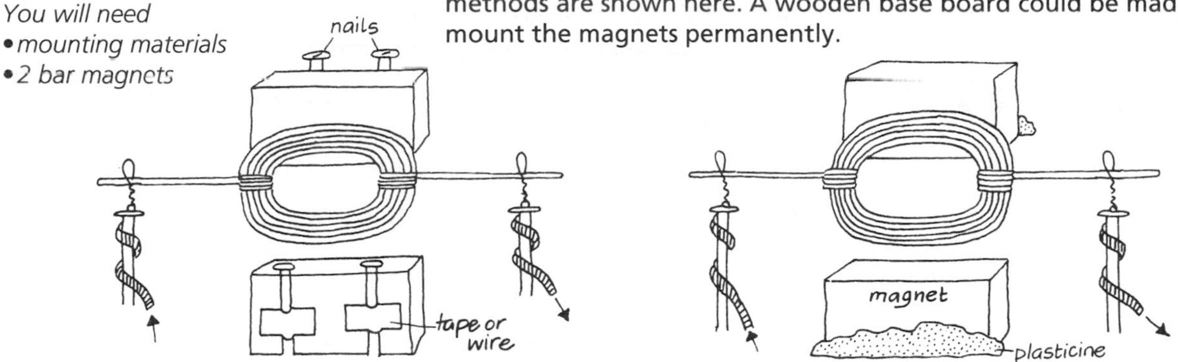

Coils

Experiment with different sizes, shapes and turns of coil. Wind a coil around a matchbox or ruler to give shape, or use a cork as shown. Note the methods for ending the coils.

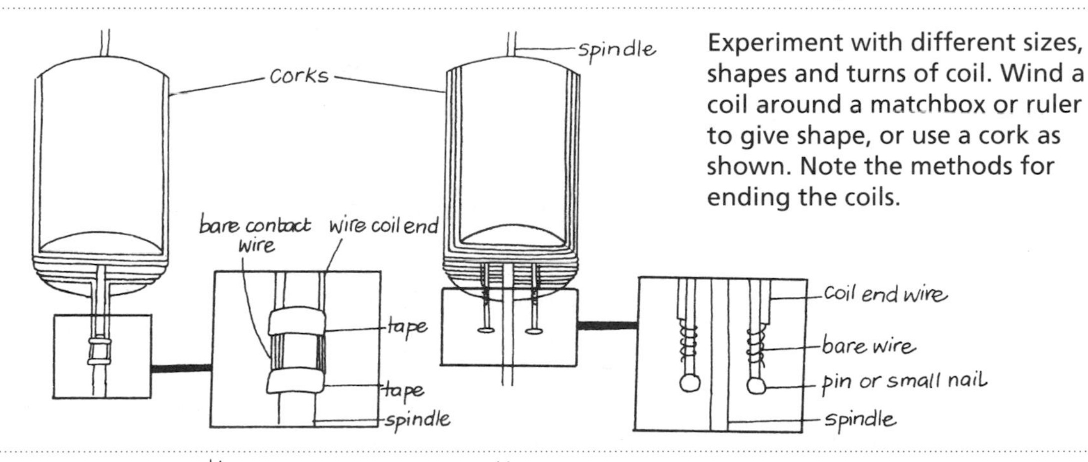

Brushes (contacts)

You will need
- *2 pins or small nails*
- *strips of thin metal*
- *fine flexible wire*

As the coil spins it will be constantly breaking and re-forming an electrical circuit. It is essential that the coil makes good contact with the brushes to keep the coil spinning. Experiment with strips of metal or fine wire as brushes. Some methods are shown.

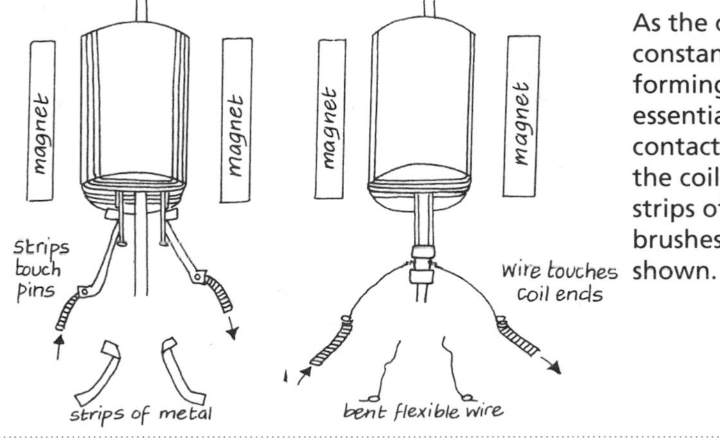

Marson Can Motor

You will need
- *2 bar magnets*
- *2 pencils or sticks*
- *wire for coil supports*
- *fine wire for coil and circuit*
- *dry cell*
- *elastic band*

Make coil supports by twisting wire around a pencil to give the shape of the hole. Make the coil by shaping it around the dry cell. If the coil wire is insulated, remove the plastic layer at each end so it runs smoothly in the supports. Attach the magnets and coil supports as shown. If the coil does not spin well, try reversing the direction of the current.

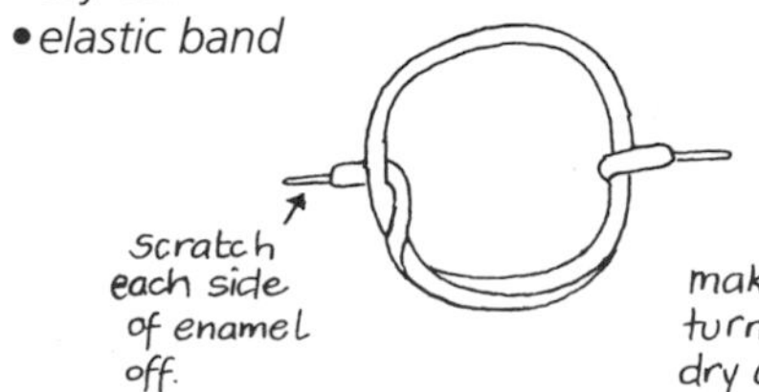

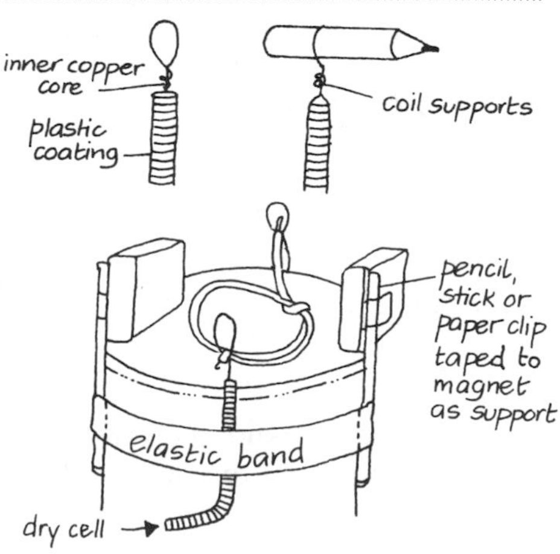

Energy forms and transducers

- **Energy** can be **transduced** (changed) from one form to another.
- **Energy sources** can be grouped into different types, e.g. **potential, kinetic, chemical, heat, sound** and **electrical.**
- Energy from the Sun can be captured and transduced into other useful forms of energy.
- Fossil fuels are a **non-renewable** source of energy.
- **Sustainable sources** (often called alternative energy sources) of energy, e.g. energy from the Sun or wind, are never used up.

Different forms of energy

Potential energy

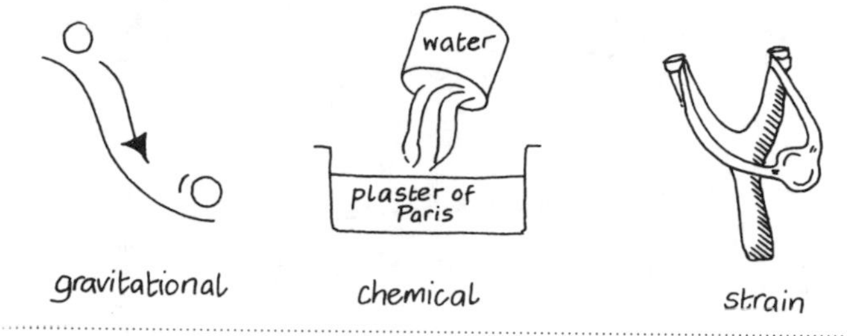

Kinetic energy

The ball and the wind both have kinetic energy.

Energy changes

Rolling tin

You will need
- *tin can*
- *rubber band*
- *weight*

Punch 2 holes in each end of a tin. Loosely fit a rubber band between the ends. Hang a weight on the band. The tin will continue to roll back and forth after it is pushed.

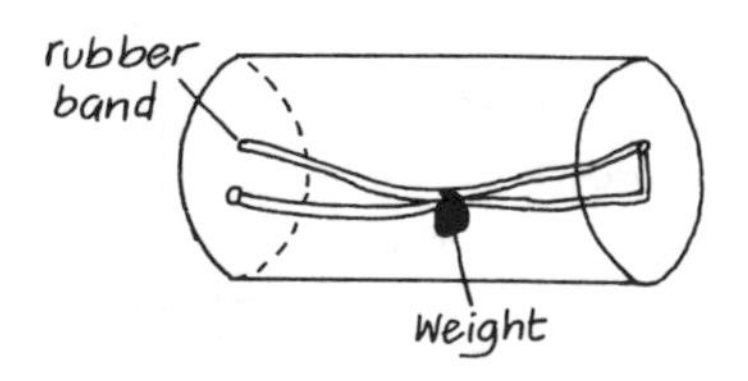

Wind turbine

You will need
- *drinking straw*
- *mounted needle*

***Safety:* Do not allow students to share blow pipes because of the danger of cross-infection.**

Cut the straw as shown. Stick the straw on the needle and ensure it spins easily. Either hold the straw in the wind or provide the wind yourself by blowing with a blow pipe.

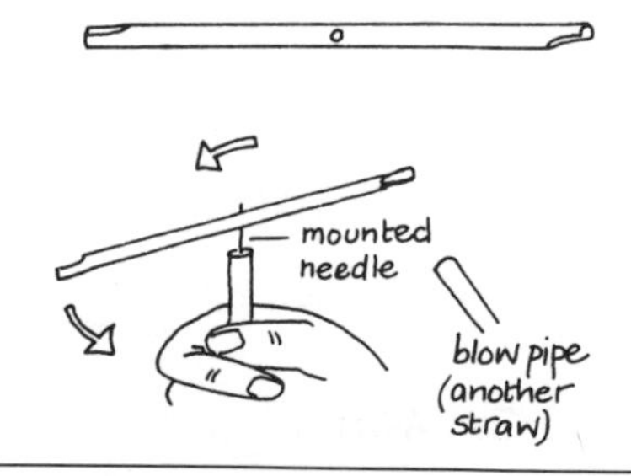

Steam engine

You will need
- *tin with a tightly-fitting lid*
- *burner or candle*
- *water*
- *light bulb*
- *card*

***Safety:* Ensure the safety bung is not too tight and that the tin is not filled with water.**

Make 3 card cones and stick them onto the light bulb. Mount the light bulb so it can spin freely. Pierce a small hole in the side of the tin. Half fill the tin with boiling water and ensure the steam comes out of the hole in a jet. Do the experiment indoors to avoid any wind.

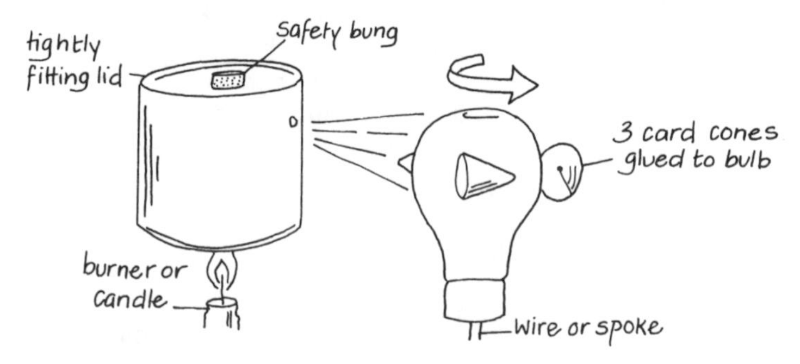

Sustainable sources of energy

Water power

The kinetic energy from water at a waterfall can turn a turbine and so generate electricity. The water can also drive a wheel and so machinery.

Wind and wave power

Windmills and wind turbines can harness wind power. A float bobbing up and down on waves can be used to drive a turbine to generate electricity.

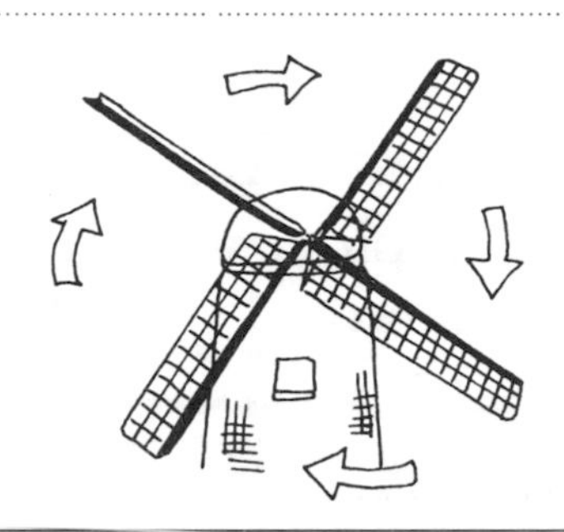

Energy from the Sun

Using tins

You will need
- *2 tins with lids*
- *black paint*
- *water*
- *sunshine!*

Paint one tin black, the other should be white or shiny metal. Put equal amounts of water at the same temperature into both tins. Put the lids on and leave them in the sunshine. After 15 minutes compare the temperatures.

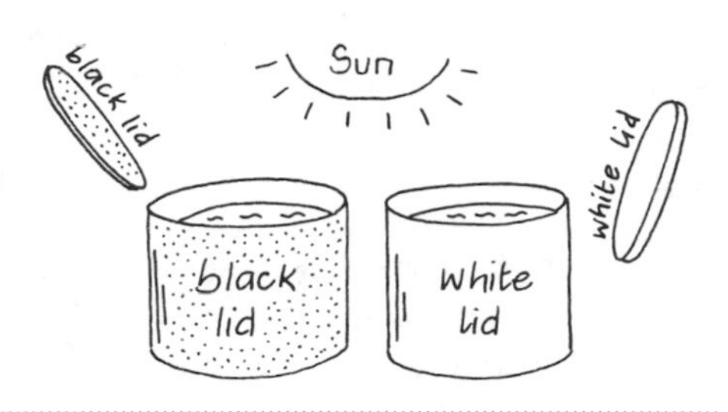

Using matchboxes

You will need
- *2 matchboxes*
- *silver and black paper*
- *2 thermometers*

Put the thermometers inside the 2 matchboxes and leave them in the sunshine. The temperature inside the black box will be higher than inside the silver box. This is because the dark colour absorbs the heat of the Sun, the silver reflects it.

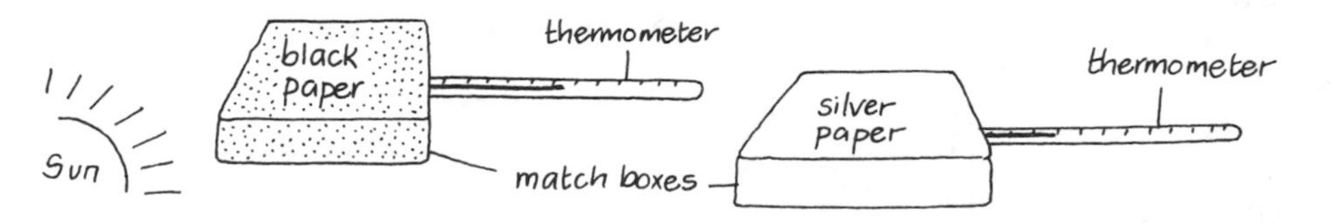

Uses of the Sun's energy

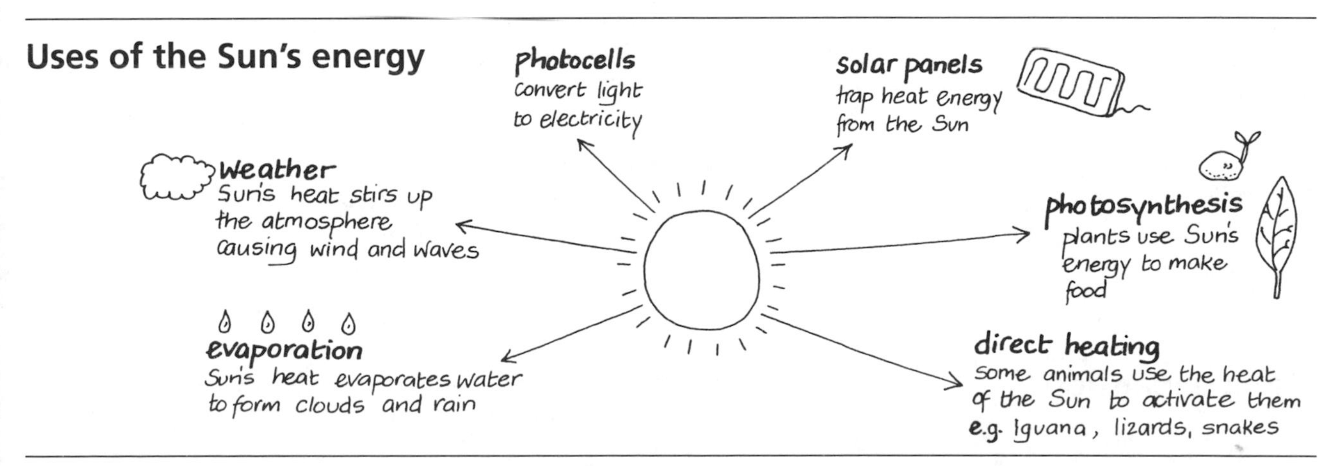

The greenhouse effect

You will need
- *2 cups of water*
- *large glass container with lid*

Put the same amount of water into both cups. Put one cup under the glass container. This is now a mini-greenhouse. Leave both cups in the sunshine for 1 hour and then measure the temperature of the water in both. The cup in the 'greenhouse' will be hotter. The Sun's heat is trapped by the glass. Burning fossil fuels produces gases such as carbon dioxide which form a layer around the Earth. They trap the heat of the Sun and so act like a greenhouse. This is called the greenhouse effect.

Heat and expansion

- **Temperature** is a measure of how hot an object is.
- **Heat** is a form of **energy** and measured in joules or calories.
- When solids, liquids and gases are heated they expand. On cooling they shrink to their original size.
- The same amount of heat applied to different materials will not make the materials both expand by the same amount. Materials have different **coefficients of expansion**.

Making a spirit thermometer

You will need

- *glass tube 5 mm external diameter and 1 mm bore*
- *hot flame*
- *coloured alcohol*
- *card*
- *dish of cold water*
- *safety goggles*

Safety: Do not heat the bubble while the tube is submerged in alcohol. Do not heat alcohol.

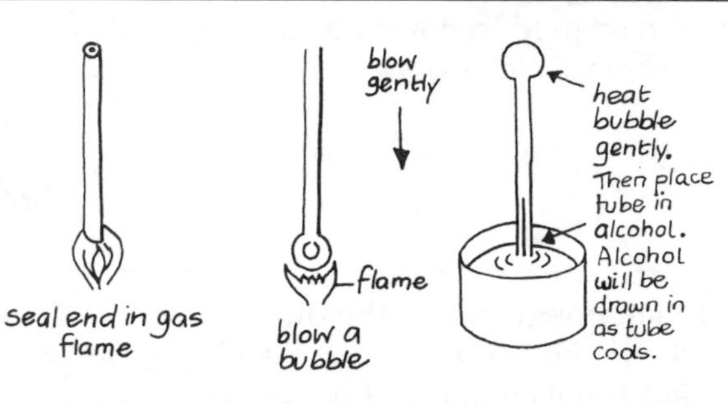

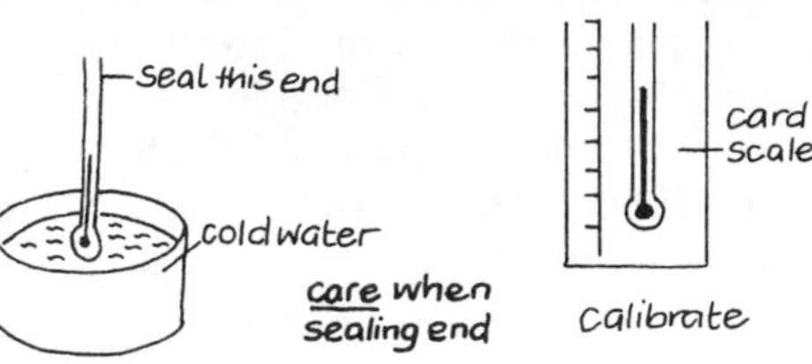

Follow the steps shown.

Calibrate the thermometer using the following:

- boiling water 100 °C
- boiling ethanol 78 °C
- body temperature 37 °C.

Draw a scale on the card or scratch it onto the tube (thermometer) itself.

Note: these calibrations apply only if you are working at, or near, sea level.

Expansion of metals

Metal between nails

You will need

- *metal strip*
- *metal disc, e.g. coin*
- *2 nails*
- *mounting board*
- *hot flame*

Place the coin between the nails, then heat the nails. The coin cannot now be removed as the nails have expanded.

Cut the metal strip so it fits exactly between the 2 nails. Heat the strip and then try to fit it between the nails.

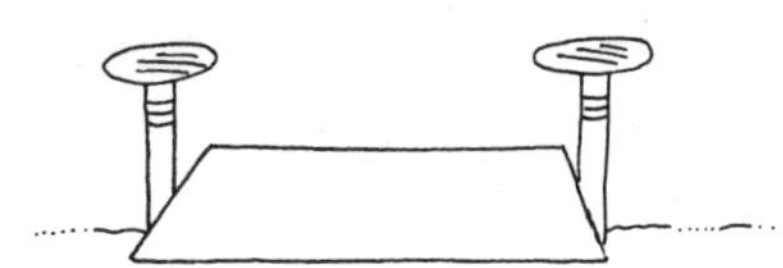

Ring and nail

You will need

- *nail*
- *wire*
- *hot flame*

Make a wire loop which is just big enough to pass over the head of the nail. Heat the nail.

Ask students why the loop will not fit over the hot nail head.

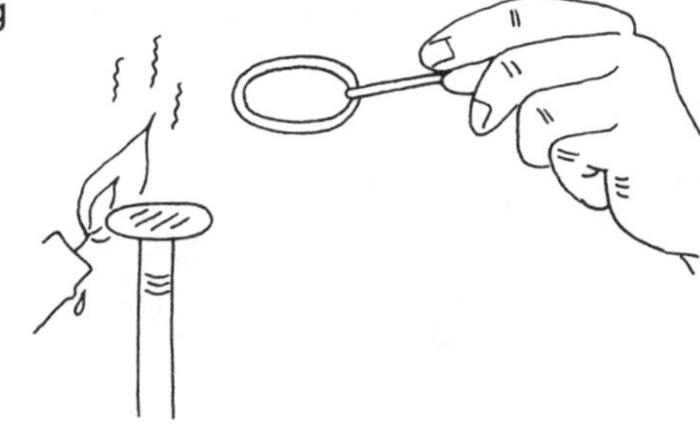

Using expansion

Stiff lids can be removed from jars with metal lids by carefully heating only the lid, e.g. in very hot water. Try the method on stiff nuts and bolts.

Problems with expansion

If boiling water is poured into glass containers the glass may split. This is because the glass on the inside expands rapidly and the outside does not. The stress cracks the glass.

Allowing for expansion

Steam and oil pipelines in hot countries often have loops to allow for expansion and contraction.

Slabs on a concrete road have gaps between them to allow the slabs to expand in the heat. Tar is put into the gaps because it is flexible.

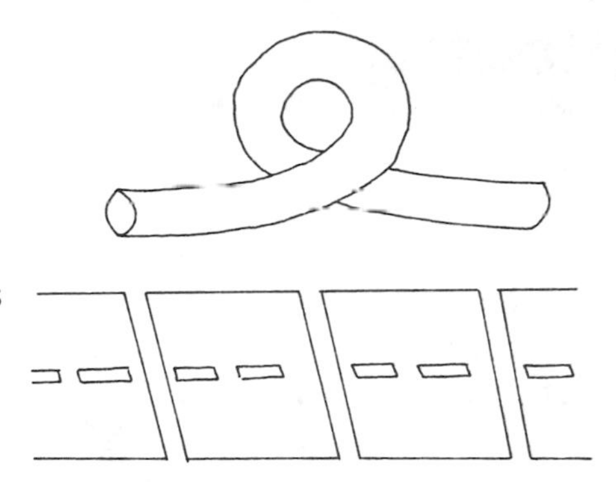

Measuring expansion

You will need
- *two bottles*
- *a cork*
- *bicycle spoke*
- *candle*
- *toothpick or round stick*
- *paper*

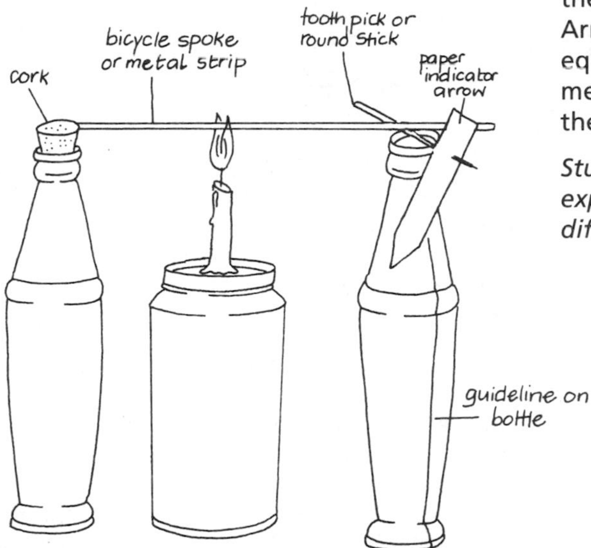

Push the spoke or metal strip into the cork so it is held firmly. Arrange the rest of the equipment as shown. As the metal is heated it expands and the indicator moves.

Students could compare the expansion of different metals and different thicknesses of metal.

Expansion of liquids

Bottle fountain

You will need
- *bottle made of thin glass*
- *bung with central tube, e.g. biro casing*

The container must be completely full of liquid, so a little of it is just visible at the bottom of the tube. When the bottle is held tightly, heat from the hands makes the liquid rise up the tube, indicating the liquid has expanded.

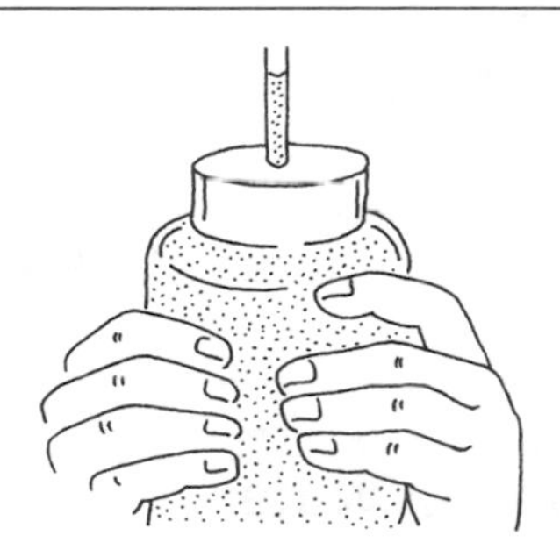

Liquid thermometers

The mercury or alcohol expands and contracts according to its temperature.

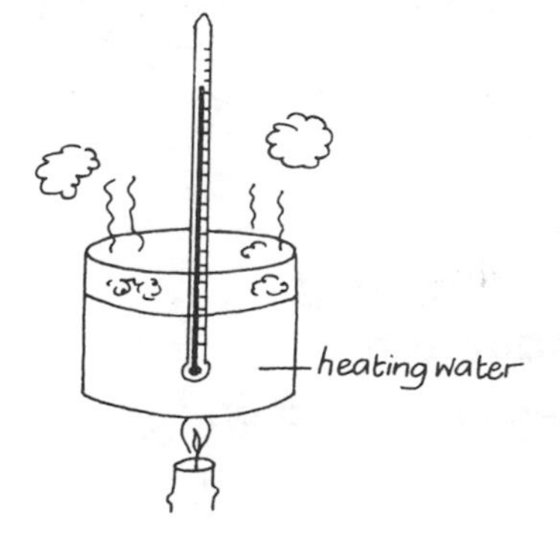

Heat

- Heat moves by **convection, conduction** and **radiation**.
- In convection and conduction heat is carried by **molecules**, e.g. of water or air.
- Radiant heat is a form of **electromagnetic radiation**, it is **infrared**.
- Radiant heat can move through a **vacuum**, whereas conduction and convection need a **medium** (water, metal, etc.).

Convection in air

Convection detectors

You will need
- *paper*
- *tin foil*
- *wire*
- *string*
- *candle*

Make the convection detectors illustrated. If they are held above a candle they will turn round.

Ask students to make their own designs.

Ventilation system

You will need
- *glass-fronted box*
- *2 cardboard tubes*
- *candle*
- *smoking rag or paper*

Make 2 holes in the top of the box and push in the cardboard tubes. Place the candle under one of the tubes. When the candle is lit, smoke will be drawn into the other tube.

Discuss with students how this principle might be used to ventilate a room, or to draw in cool air to a container.

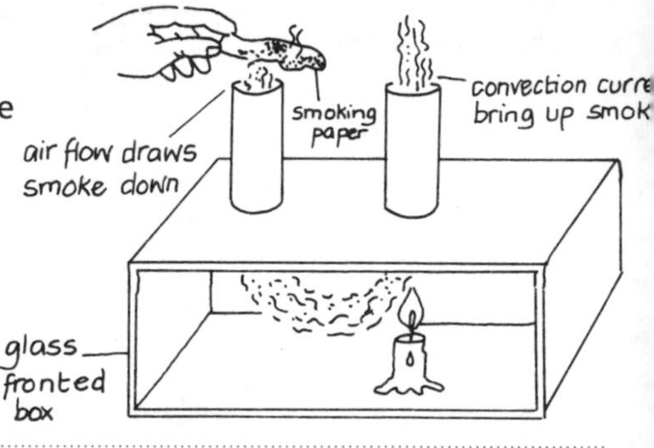

Hot air balloons

You will need
- *lightweight paper bag*
- *candle*
- *cardboard tube*

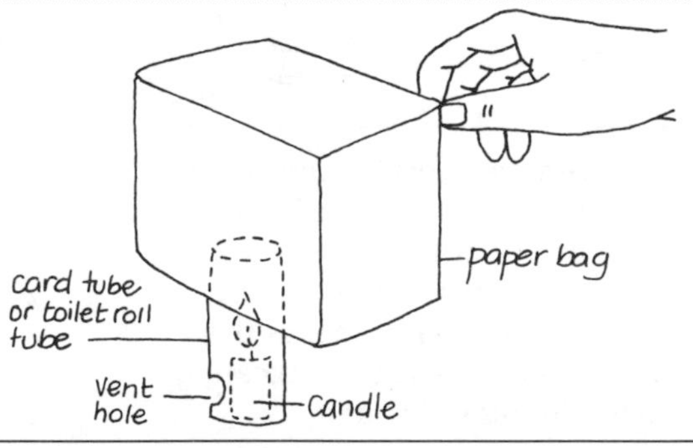

If the bag is held over the candle it will rise as the air inside heats up. This is because warm air is lighter than cool air.

Students could design their own hot air balloons and test which flies highest.

Ask students why the top of the oven is the hottest part.

Convection in water

You will need
- *sawdust*
- *glass container*
- *candle*

Put water and a small amount of sawdust in the container. As the container is heated the convection currents will be visible.

Discuss with students the application of this principle to domestic hot-water systems, and to land and sea breezes.

Studying heat conduction

Speed of conduction

You will need
- *candle*
- *metal rod*
- *small stones, pins or seeds*

Use candle wax to stick small stones onto the metal rod at regular intervals. Put a cloth or handle (e.g. maize core) around one end of the rod. When the rod is held in the flame the stones will drop off as that part of the rod gets hot.

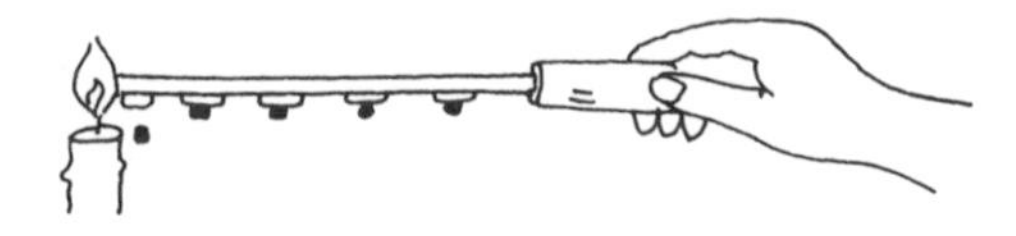

Measuring conduction rates

You will need
- *wires or rods of different metals*
- *candle*
- *small stones, pins or seeds*

Stick small stones along the wires using candle wax. Hold the wires in the flame and record the time each of the stones drops off each rod/wire, i.e. the speed of conduction.

Results could be recorded on a graph.

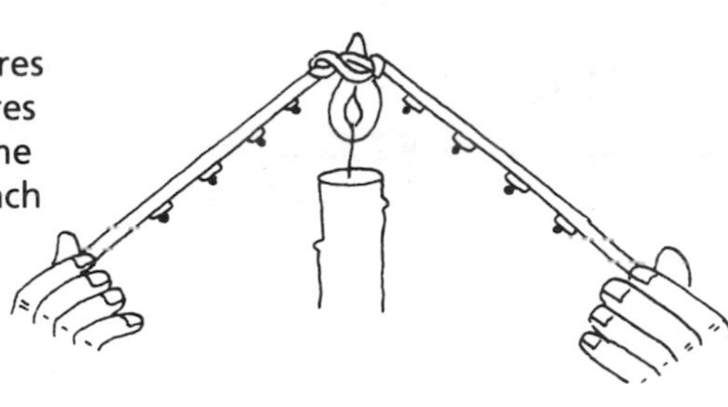

Radiation

Observing radiation

Glass windows block any movement of air. Therefore the heat is carried into a classroom by radiation.

Focussing radiation

You will need
- *convex lens*
- *paper*

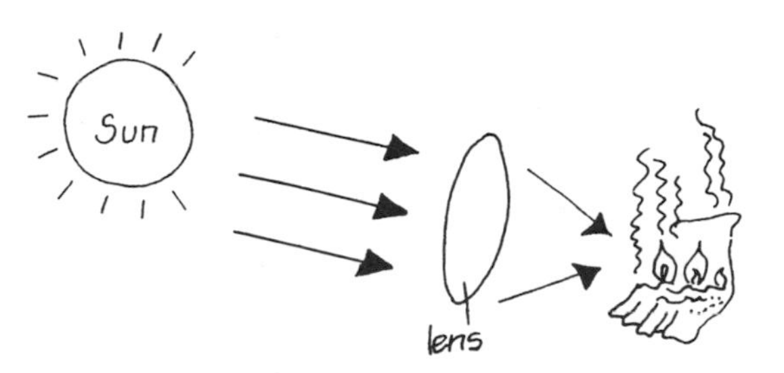

Focus the rays of the Sun onto paper until it catches fire.

As an extension, discuss how this is used in a solar oven.

Radiation from surfaces

You will need
- *shiny can*
- *black can*
- *white can*
- *hot water*
- *thermometer*

Fill all 3 cans with the same volume of hot water. Cover them with lids and stand them in a cool place. Record the temperature of the water in each can every 5 minutes. Black surfaces both absorb and radiate heat more quickly than shiny or white surfaces.

Amazing tricks

Paper pan

The pan will not burn as the temperature of the paper never rises above 100 °C.

Non-burning paper

A coin on a piece of paper conducts heat away before the paper burns.

Fireproof material

A coin conducts heat away before the cloth can burn. Do not use synthetic materials as many melt at quite low temperatures!

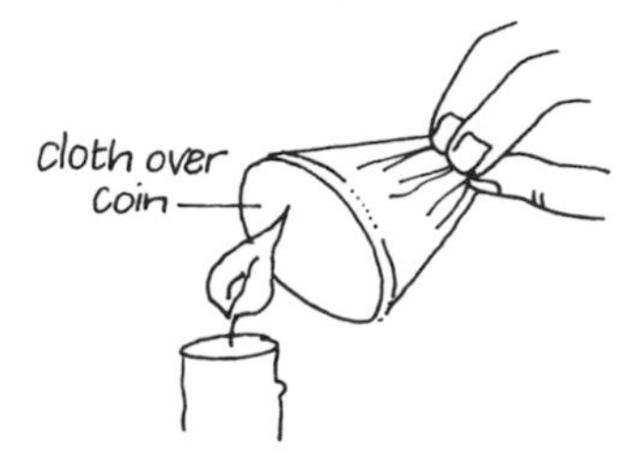

Force and motion

- A **force** is a push or a pull exerted on an object.
- For every **action** (force) there is a **reaction** (second force).
- If an object is stationary the action is equal and opposite to the reaction. An object moves when one force is greater than the other.
- **Friction** is the force which prevents objects sliding over each other.
- Reducing friction reduces the size of force needed to move the object, i.e. it takes less **energy** to move the object.

Examples of forces

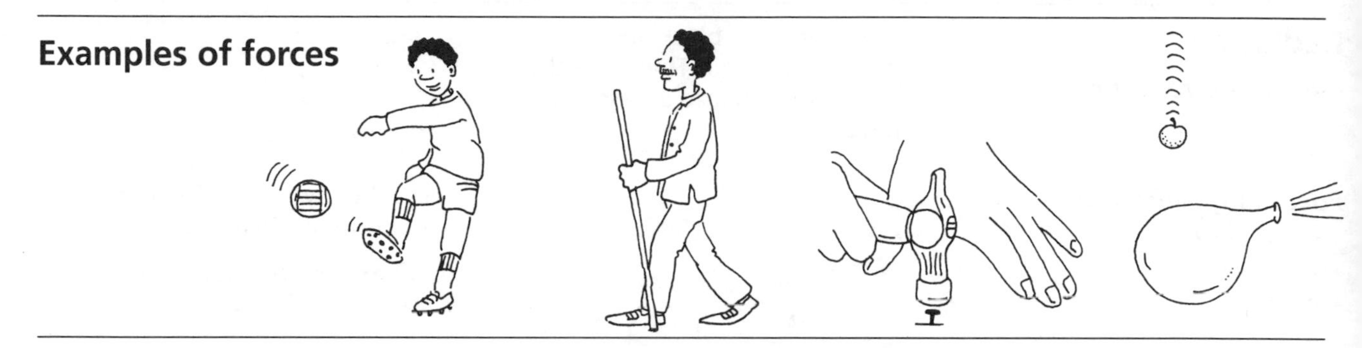

Action and reaction

Jet balloon

You will need

- *string*
- *straw*
- *tape*
- *balloon*

When the balloon is inflated and released it moves.

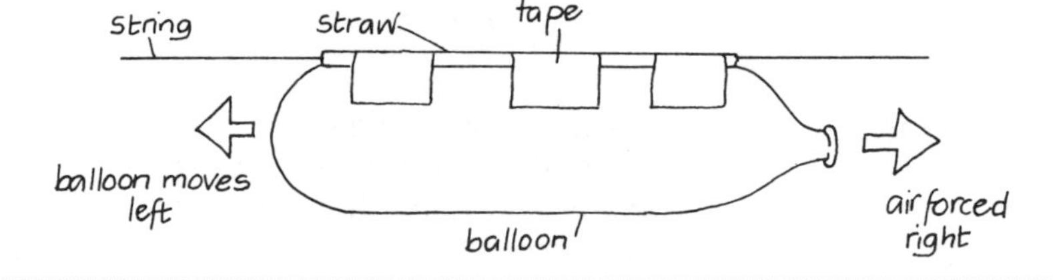

Other examples

A jet airliner throws out hot gases from its engines in one direction (action), called the thrust. The plane moves in the opposite direction (reaction). The canoe shown here also moves away from the push action.

Bridges

Forces which weaken bridges

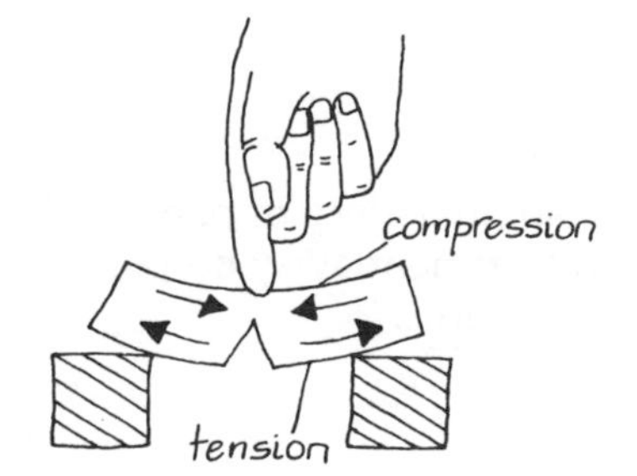

The bridge bends under the weight of the load. More than one force is at work. Compression forces are concentrated on the top surface. When a bridge bends, compression on top creates tension forces on the bottom surface.

Strengthening bridges

You will need

- *books*
- *string*
- *base board*
- *nails*

In a suspension bridge the tension in the bridge is increased by securing the 'strings' and suspending them over towers or from trees.

Ask students to build the 2 bridges shown. Discuss why the suspension bridge is stronger.

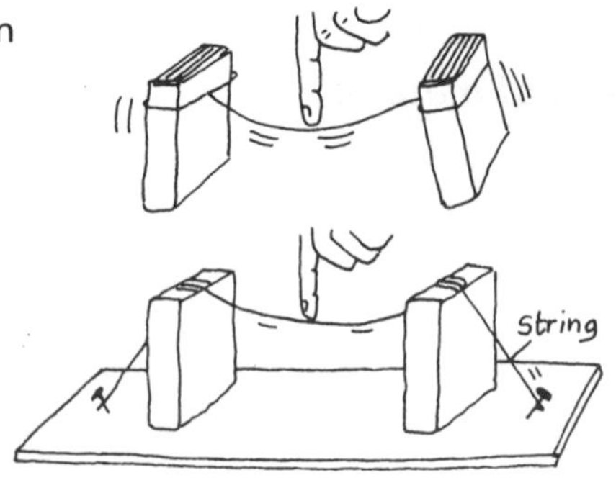

Friction

Useful friction

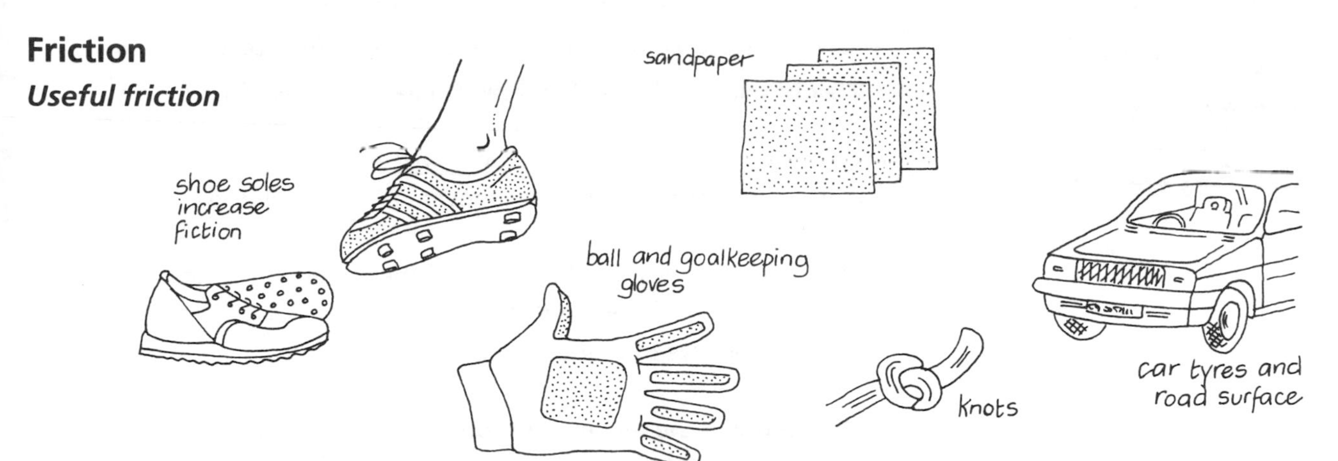

Problems with friction

Friction causes

- bald tyres on cars
- wear in car engine parts
- wear in rope bridges
- soles of shoes to wear out.

Reducing friction

Model truck

You will need

- *brick or heavy book*
- *pencils or marbles*

Pencils and marbles are both rollers and so reduce friction.

As an extension use a force meter to measure and compare frictional forces using different loads and identify ways of reducing friction.

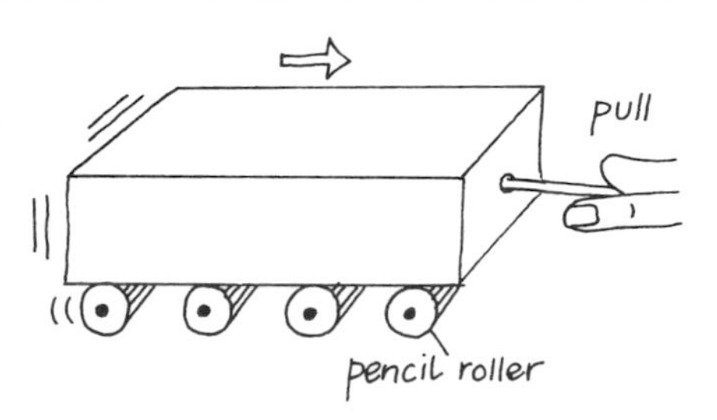

Ball bearings and oil

You will need

- *marbles*
- *oil*
- *2 tins*

First test the effect of oil by putting a little between finger and thumb and noting how they slip over each other.

Put oil onto the marbles and note how easily the top tin turns on the lower one. The rolling of the balls and the lubrication of the oil both reduce friction.The marbles are acting as ball bearings. Many engines have ball bearings.

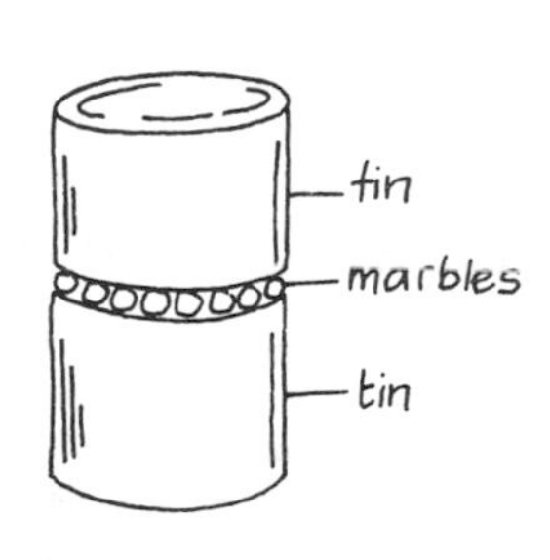

Water as a lubricant

You will need

- *a sheet of glass*
- *glass tumbler*
- *water from a tap*

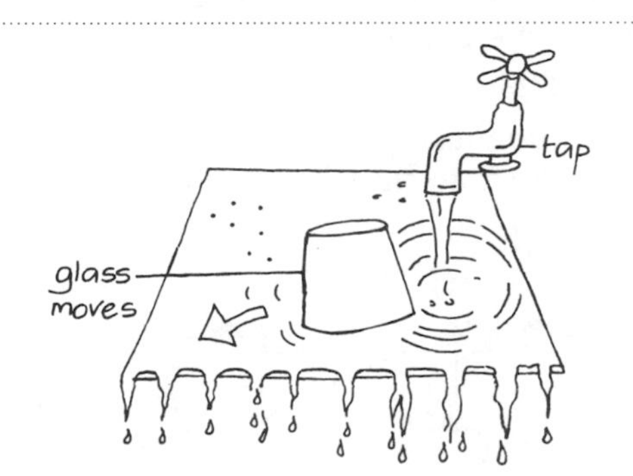

Leave a little water in the glass before inverting it. The glass floats on a cushion of air and water like a hovercraft.

Force and motion *continued*

- **Force** is measured by a force meter. The unit of force is a **newton**.
- **Weight = mass** x **force of gravity**. On Earth a mass of 1 kg has a weight of 1 kg, but on the Moon a mass of 1 kg would *weigh* less than 1 kg because the force of gravity is less.
- **Inertia** is the tendency to resist movement. It depends only on mass, not on weight.
- **Motion** is movement and can be described in different ways, e.g. speed, velocity, **acceleration**.
- **Speed** is the rate of change of position; **velocity** is speed in one direction.
- **Momentum** is the tendency to continue moving in the same direction. It can be passed on from one body to another. Momentum = mass x velocity.

Making a force meter

You will need
- *drinking straw*
- *wire*
- *steel spring*

Make the spring by winding the wire around a rod of the same diameter as the straw. Alternatively make the spring by attaching one end of the wire to a drill and twisting it. Make the diameter of the spring slightly smaller at one end so it will grip the straw. Calibrate the spring by using known weights.

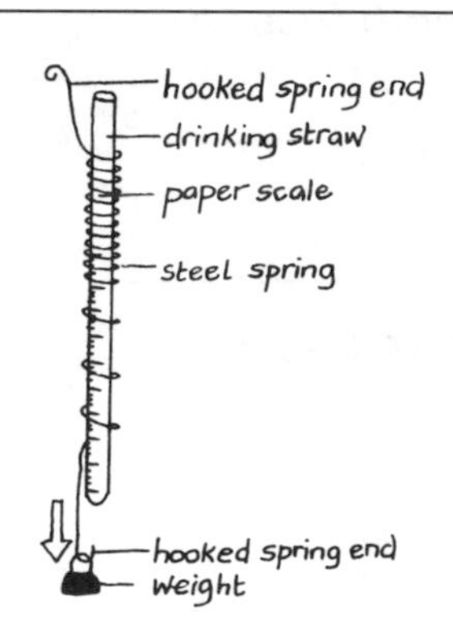

Centripetal and centrifugal forces

Spinning forces

You will need
- *string*
- *2 weights*
- *ballpoint casing*
- *paper clip*
- *scissors*

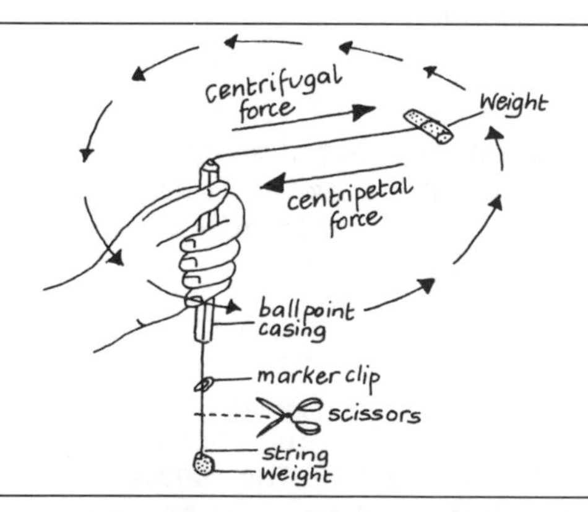

Put the casing onto the string and attach a weight to each end of the string. Note what happens to the clip if the weight is spun faster. Cut the string in midspin and note what happens.

Inertia and momentum

Tin can pendulums

You will need
- *2 buckets*
- *long strings or rope*
- *water, sand or stones*

Hang each of the buckets from a long string. Inertia has to be overcome to start a bucket swinging.

Ask students which bucket needs most force to start it swinging and why.

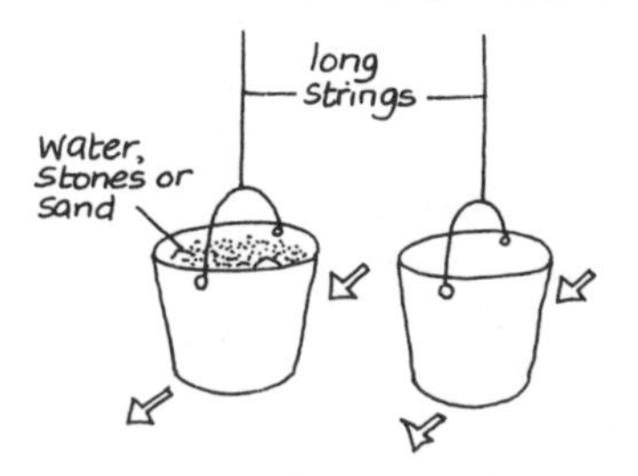

Bumping bottles

You will need
- *wooden support*
- *string*
- *2 bottles*

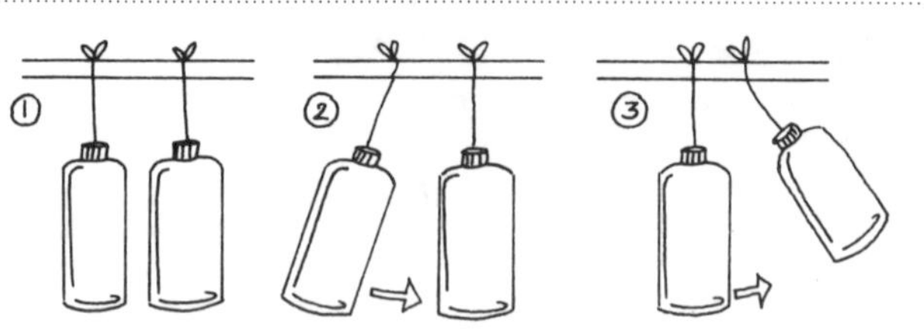

The bottles are smooth and hardly touch as the momentum is transferred from one bottle to the other. Momentum is conserved.

Amazing examples

Coin and paper

You will need
- *coin*
- *thin card*

Flick the card away. The inertia keeps the coin in place. Try the same trick with the card over a jar or glass. A heavy coin works best.

Pile of books or coins

You will need
- *pile of coins*
- *pile of books*

Try to remove the bottom book without upsetting the pile. Impossible? To remove the bottom coin from a pile, flick another coin at it. The momentum of the flicked coin is transferred to the bottom of the pile. The momentum overcomes inertia.

Dropping fruit

You will need
- *apple or similar fruit*
- *knife*

The farther the fall, the greater the momentum and the deeper the cut.

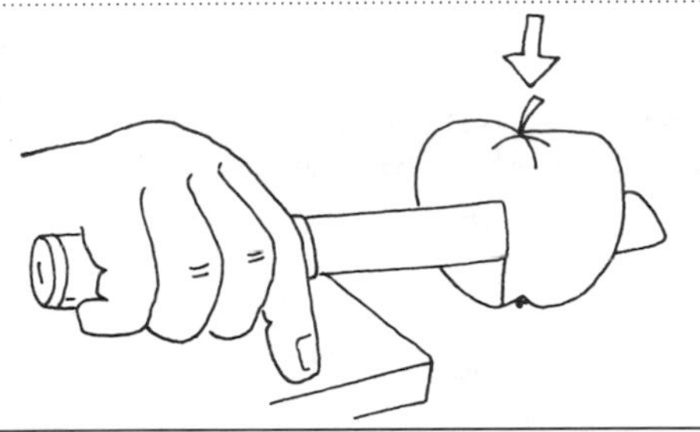

Measuring motion

Making the vehicle

You will need
- *block of wood*
- *bottle or box*
- *cotton reels or bottletops*
- *wire*
- *matchboxes*
- *sand*

Make the wheels using, e.g. bottle tops, cotton reels. Attach these to the wooden base. Attach the box or bottle to the wooden base. Make standard weights by adding the same amount of sand to each of the matchboxes.

A child's toy car would be a ready-made alternative.

Making the timing cup

You will need
- *container*
- *dilute ink*
- *cotton string*
- *pin*

Pierce a small hole in the bottom of the cup and seal it with the pin. Fill the cup with ink. When the pin is pulled out the ink will fall from the cup in regular drops. You will need to experiment a bit so the drops fall regularly and fast enough.

Test slope

You will need
- *long, thin strips of paper*
- *pile of books*
- *board*

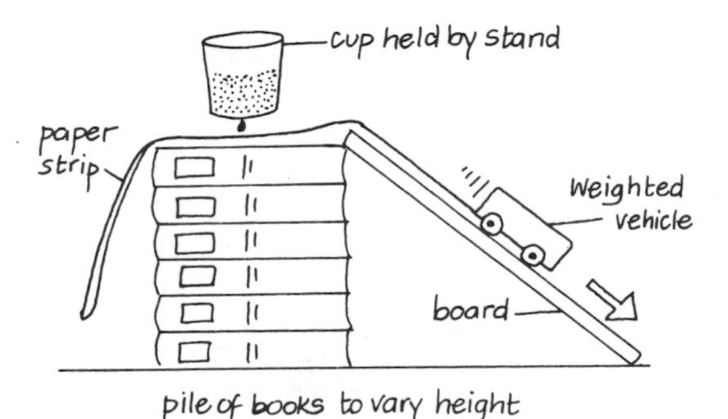

Pile the books to make slopes of different heights. Attach the ticker tape to the weighted vehicle. When the vehicle is released, pull out the timer cup string. Make sure the ink does not damage the books.

Ask students to measure the velocity of the vehicle while altering either the loads in it or the height and angle of the test slope. This could be done qualitatively and results recorded on graphs.

Levers, pulleys and machines

- A **machine** allows a **force** applied in one place to overcome a force applied in another place. Levers and pulleys do **work** and are simple machines.
- A **lever** transmits a force from one place to another.
- The **principle of moments** is used to calculate the relationship between **load, effort** and the distance of each from the **fulcrum**.
- **Pulleys** change the direction of a force.
- In a system of pulleys, if the effort moves a long way and the load only a short distance, then a small effort can lift a heavy load.

Moments

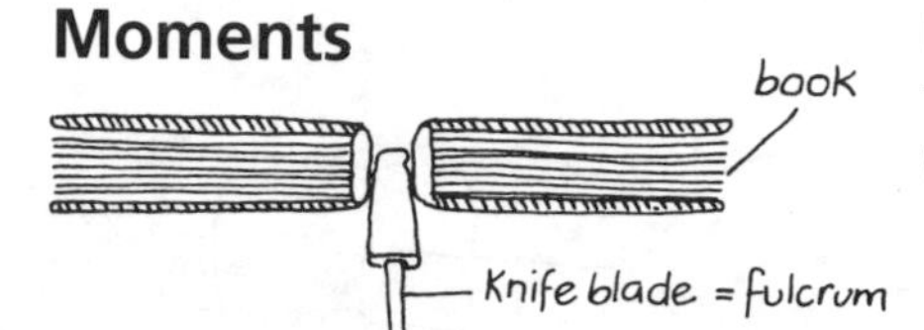

In order to balance around a fulcrum the forces on either side must be equal.

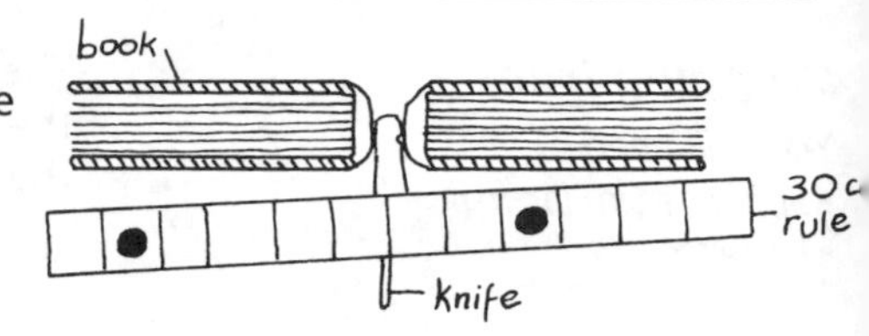

A lever balance

You will need
- *ruler or measuring stick*
- *weights, e.g. coins*
- *fulcrum, e.g. knife*

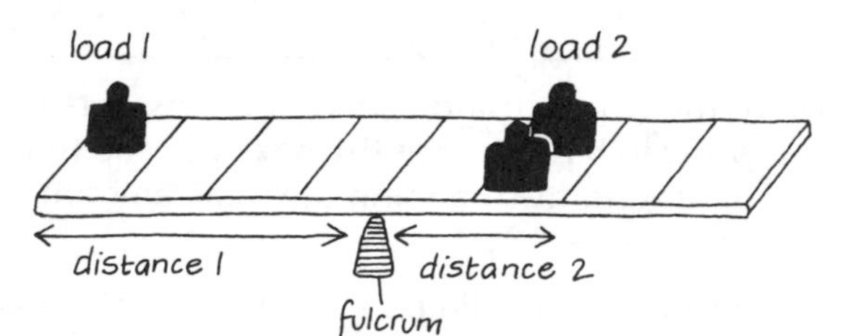

Balance the weights on the ruler. Investigate the relationship between the size of a weight and the distance from the fulcrum on either side of the fulcrum (balancing point). Students should discover that if distance 1 x load 1 = distance 2 x load 2 the lever will balance.

Levers

Doors as levers

You will need
- *a door*
- *several hooks*
- *string*

Place the hooks in the door 10–15 cm apart. Attach the string to the hooks, one at a time. Try to pull the door open.

Ask students why it is easier to open the door if the string is far from the hinge. Investigate whether the length of string affects the effort required.

Using levers

Levers can reduce the work needed to move loads.

Ask students where levers are used in their own communities.

Water from a well

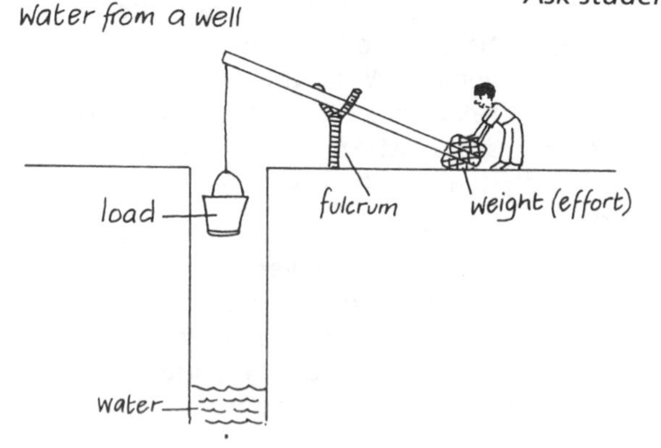

pulling a vehicle out of mud

stick (lever)

stone (fulcrum)

Pulleys

Simple pulleys

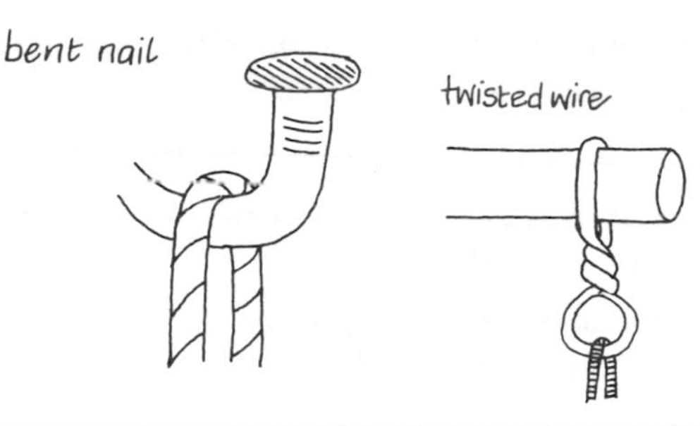

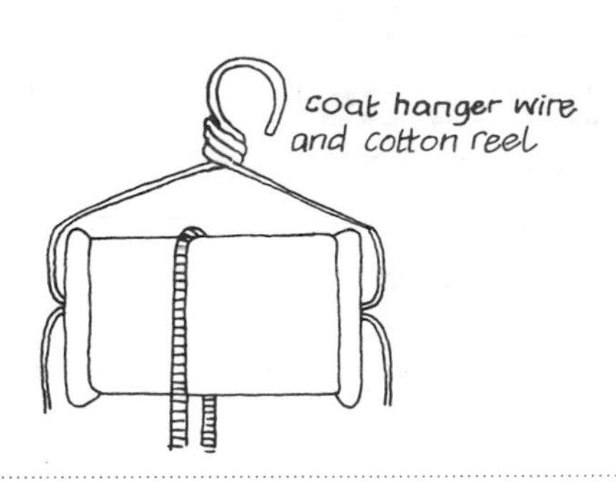

Making a pulley and holder

You will need

- *2 corks*
- *2 pins or nails*
- *strong glue*
- *2 metal strips, e.g. packing strips*
- *wood support*
- *2 washers*

Shape the corks slightly so the string runs smoothly. Drill holes into the metal strips and attach them to the wooden support. Washers will reduce friction and so aid turning.

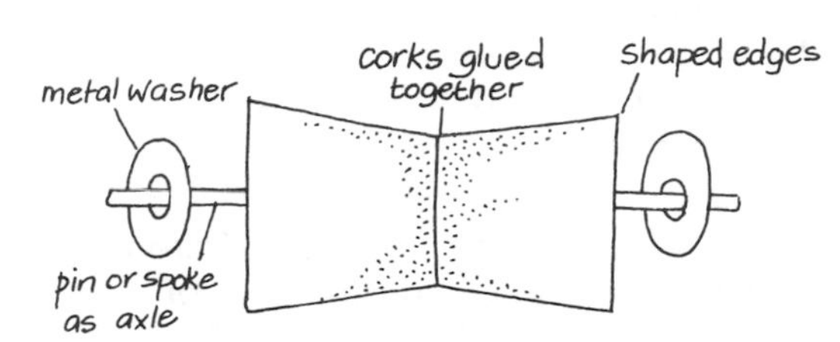

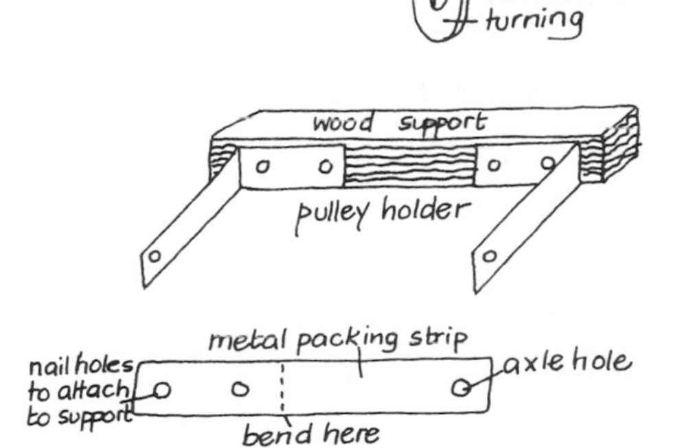

A pulley system

You will need

- *2 cotton reels*
- *long string*
- *strong wire, e.g. coathanger wire*
- *support*
- *strong nail*
- *weights*

Make the pulley system shown. Weights can be put into a plastic bag which is then hung from the hook. Accurate weights can be made by filling a bag with known volumes of water (1 cm^3 weighs 1 gram).

Ask students to investigate the effort needed to lift loads of different weights using 1, 2 or 3 pulleys.

Discuss the relationship between distance moved by the effort compared to that moved by the load and also the effort required to lift the load. This could be done quantitatively using known weights for load and effort and recording on a graph.

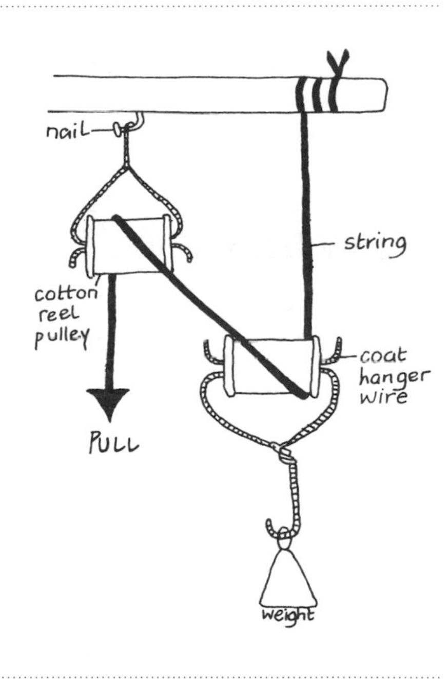

Uses of pulleys

Ask students where they have seen pulleys used and why they reduce the work of lifting loads.

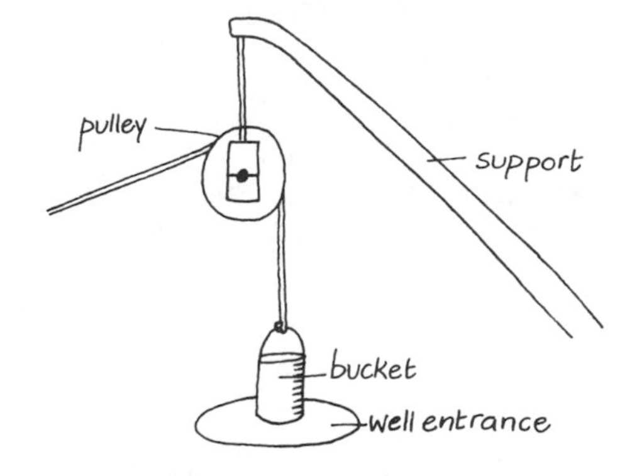

Waves as energy carriers

- The **medium** through which a **wave** travels does not itself travel with the wave.
- The **energy** of the wave moves in the direction of travel.
- In **transverse waves** the particles of the medium **vibrate** at right angles to the direction of the wave's travel.
- In **longitudinal waves** the particles of the medium travel in the same direction as the direction of travel.
- The shape and size of a wave is described by its **wavelength** and **amplitude**.

Water waves

You will need
- *cork*
- *large bowl of water*
- *weights*

Drop a weight into the water from different heights to create waves of different strengths. Observe the force of the wave by the bobbing of the cork. Alter the depth of water and note the effect on the height of the wave and the frequency of the cork's bobbing.

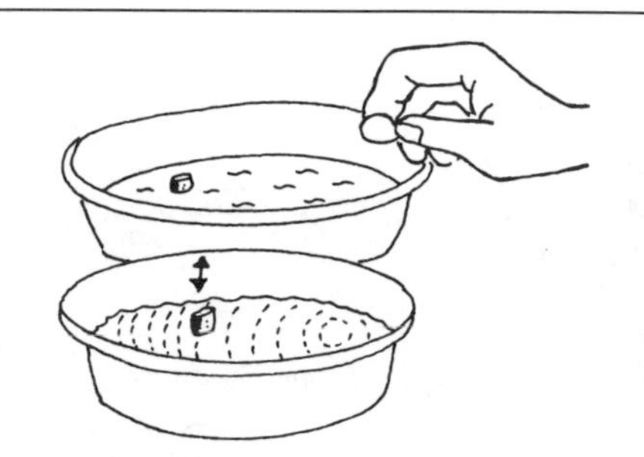

Breaking waves

You will need
- *large tray of water*
- *heavy rocks*

Tap or move the tray to create waves of different sizes. Waves will only break in shallow water, i.e. when the depth of water is less than the amplitude of the wave.

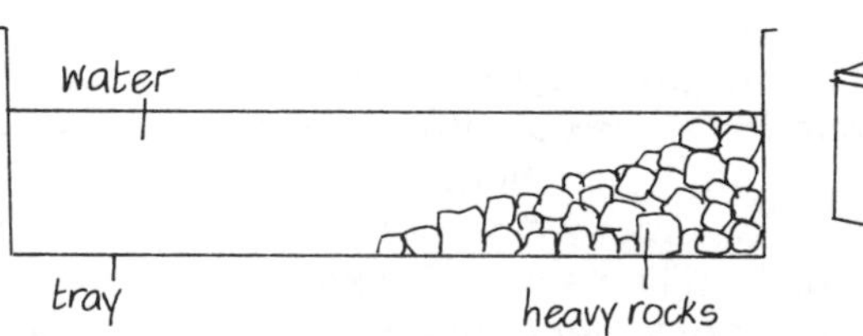

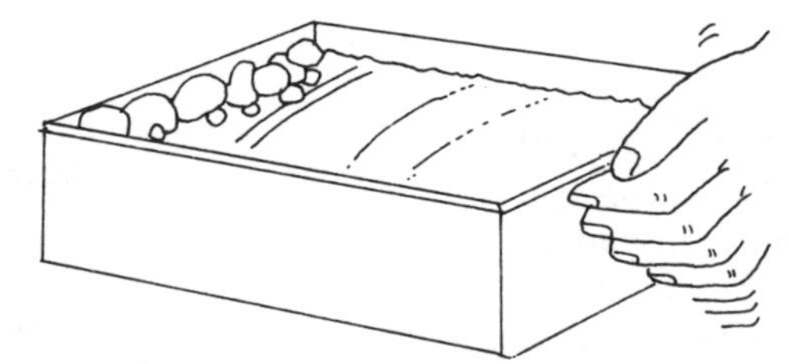

Energy from waves

The ducks, or floats, oscillate up and down on the waves. The energy from this motion is converted to electrical energy inside the duck (see page 83).

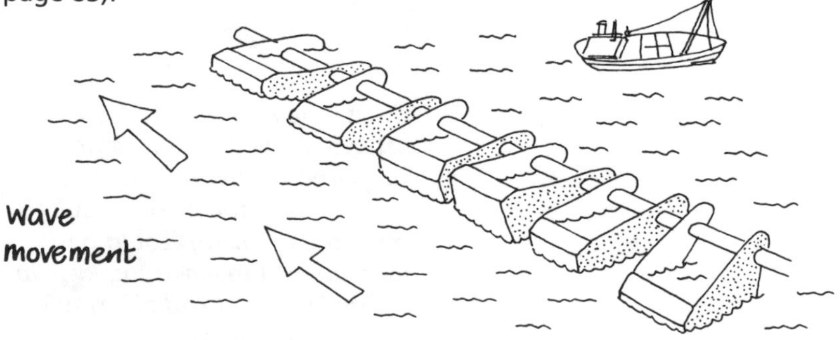

Transverse waves

Pendulums

You will need
- *weights (all identical)*
- *string*
- *support*

Arrange the weights as pendulums of different lengths as shown. They should be tied onto the horizontal string at equal distances apart. Swing one pendulum at right-angles to the horizontal string. The other pendulums will eventually swing too and the horizontal string will make waves.

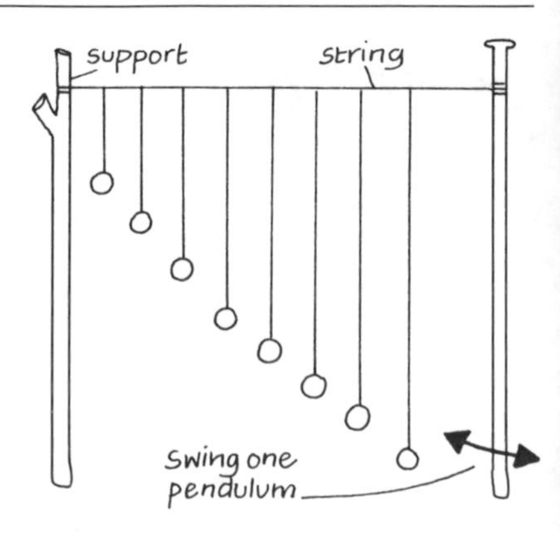

Mexican wave

The longer the line of students the more impressive this is. However, it is not easy to see the wave if you are a part of it!

Flick-sticks

You will need

- *straws or sticks*
- *rubber or paper strip or adhesive tape*
- *glue*

Cut the straws or sticks so they are all the same length. Glue them to the tape. A tape of 3 metres is a good classroom size. Adhesive tape means there is no need to glue, but unless the sticks/straws are light they may fall off. Rubber strip, e.g. from a tyre inner tube, is strong enough to last well. Twist or flick the strip to set off waves.

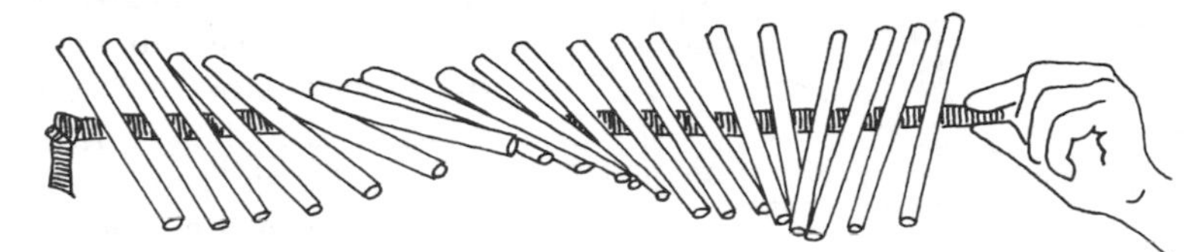

Experiment with students by altering, for example, the length of sticks or strip until really good waves can be created.

Ripple tank

You will need

- *sheet of glass*
- *wooden, plastic or glass strips*
- *waterproof glue*
- *large box*
- *mirror*
- *wooden support*
- *wire*
- *string*
- *small cork*

Glue the strips to the glass sheet (using suitable waterproof glue) to create a shallow glass-bottomed dish. Arrange the mirror in the box so that it can direct light up through the glass and project an image of the ripples on a wall. Ensure the water is still. To create circular waves dip the cork once into the water, or tap the support.

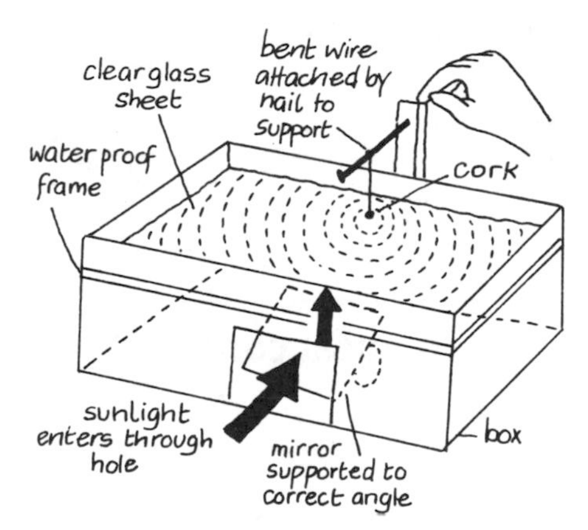

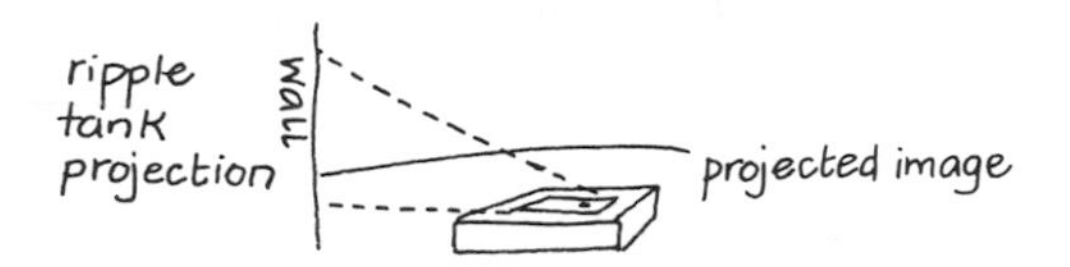

Longitudinal waves

You will need

- *marbles or coins*

A line of students can demonstrate a longitudinal wave. The student should suddenly push the student in front as soon as he/she feels the push from the person behind.

The same principle can be demonstrated with marbles or coins. (See also momentum, page 90.)

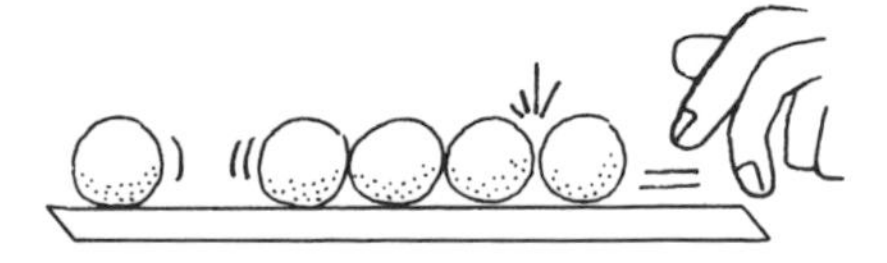

Sound

- **Sound** is caused by **vibration** of air particles.
- The **speed** at which sound travels depends on the **medium** through which it is travelling. (See page 94.)
- Loudness depends on the **amplitude** of the sound wave. The larger the amplitude the louder the sound.
- **Pitch** depends on **frequency** – the higher the frequency the higher the pitch.
- The longer the length which is vibrating, e.g. string or column of air, the lower the frequency.

Drum vibrations

You will need
- *tin can*
- *pieces of old balloon*
- *funnel*
- *candle*

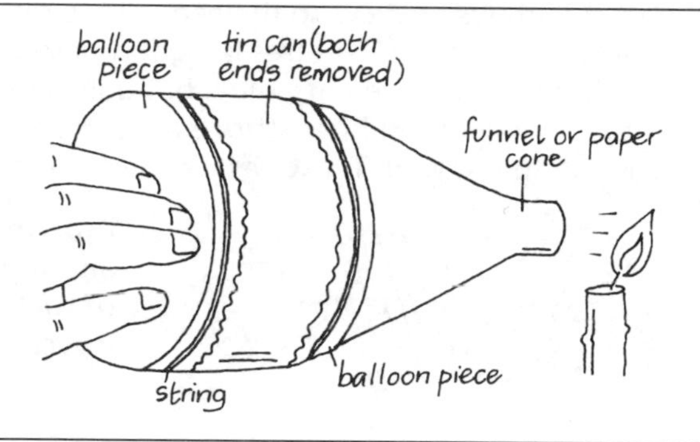

Remove the ends from the tin. Tie the balloon pieces to the tin as shown and attach the funnel. When the drum is tapped hard, sound vibration is carried through the air in the can, making the other sheet vibrate. The funnel concentrates the sound vibrations so that the air from the funnel can snuff out the candle.

Changing pitch

You will need
- *ruler*
- *table*

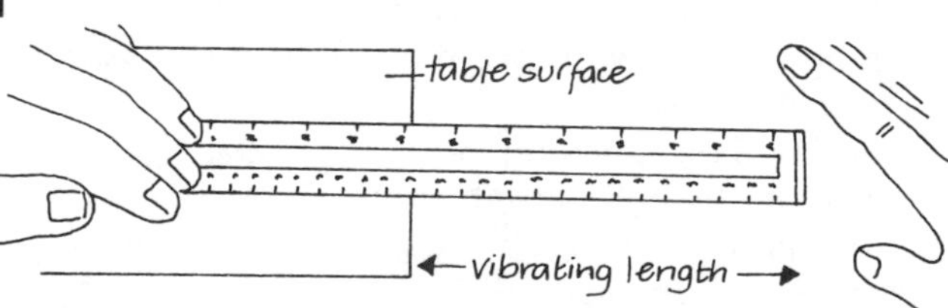

The shorter the vibrating length, the greater the number of vibrations and the higher the note.

Human vocal cords

You will need
- *rubber band*
- *2 nails*
- *wooden base or tin can*
- *pencil*

The nails should be firmly attached to the wooden base, or tin can. The rubber bands represent the vocal cords. The tighter the band, the higher the pitch.

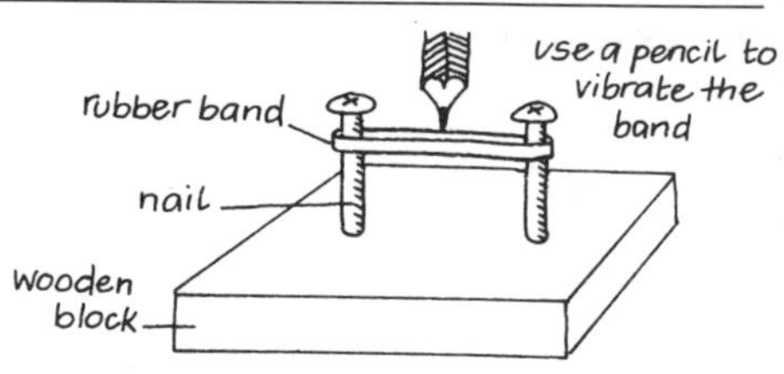

As an extension investigate the effect of using thinner bands.

Sound through solids

Tin can telephone

You will need
- *2 tin cans*
- *long piece of string*

Make a small hole through one end of each tin and remove the other ends completely. Push the string through the holes and tie 2 large knots to keep it in place. The string should fit tightly into the holes.

Investigate the following with students:
- *using wire instead of string*
- *altering the length of wire or string*
- *tautness of wire or string.*

Sound through wood

The tapping can be heard as sound travels through the wood. This effect is even more impressive if the tap is made at one end of a long piece of wood and the listener is at the other end.

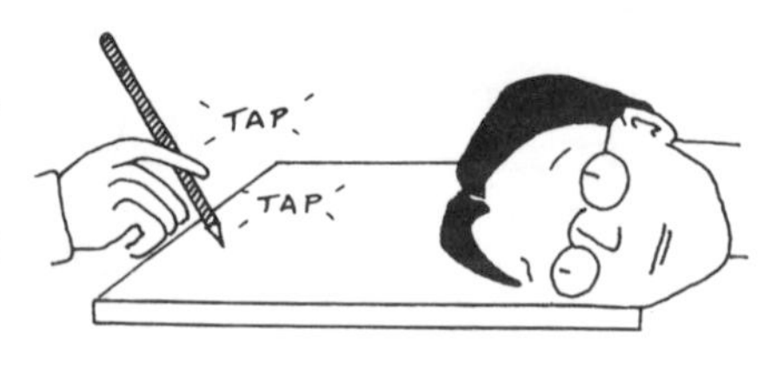

Speed of sound

Sound travels at approximately 340 m/s. Light travels at about 3 million metres per second, i.e. in effect, you see a light as soon as it is switched on. Counting the time between the flash of lightning and the thunder clap allows you to calculate how far away the storm is.

Musical instruments

All music is made of vibrations.

Ask students what local musical instruments could be used in school.

Ask how they create the vibrations.

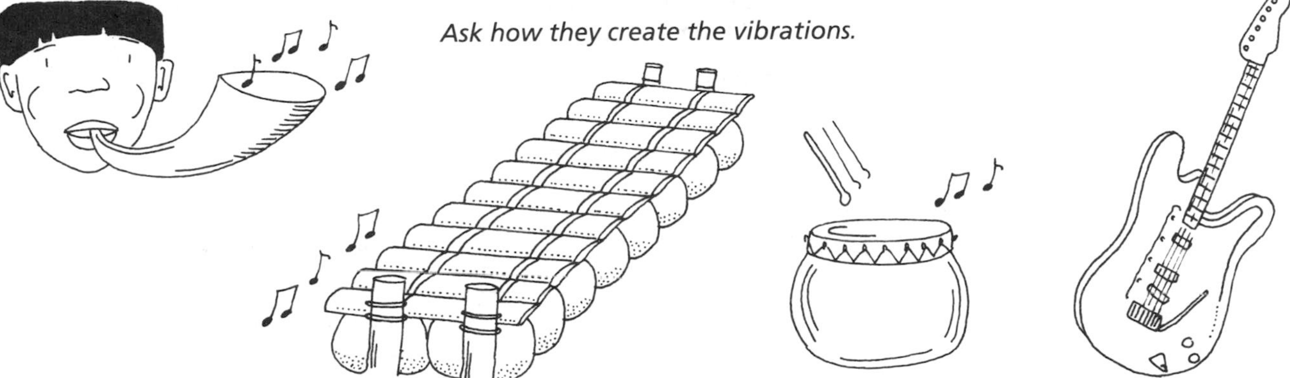

Bamboo organ

You will need
- *pieces of bamboo*
- *string or tape*

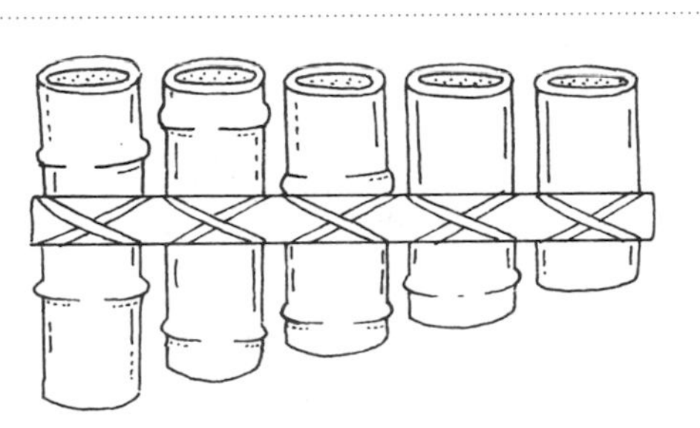

Hollow out the bamboo pieces and attach them as shown. The length of the pipes determines the pitch of the sound.

Bamboo flute

You will need
- *bamboo*
- *drill to make holes*
- *sharp knife*

Cut a wedge-shaped hole at one end of the hollow tube. This is the mouthpiece. Make pitch holes 0.5 cm apart.

sharp-edged hole

holes ½ cm apart

Hanging objects

Strike each object with the same striker.

Ask students to select objects so that a scale of notes is made.

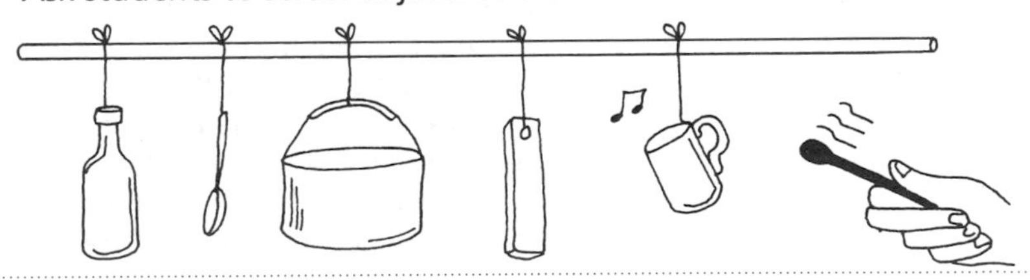

Bottle harmonica

You will need
- *glass bottles*

Put a different amount of water in each bottle. If the bottles are exactly the same size and thickness it is easier to see the relationship between pitch and volume of air above the water.

Light

- **Light** consists of **waves** of **electromagnetic radiation** and travels in straight lines.
- Light can travel through some **media**, e.g. water, glass and across a **vacuum**.
- Light may be **absorbed** by some substances (media), **reflected** or **refracted**.
- Lenses bend light, **convex** lenses bring rays together, **concave** lenses spread them apart.
- Light shining through a narrow slit will exhibit **diffraction** and **interference**.

Ray boxes

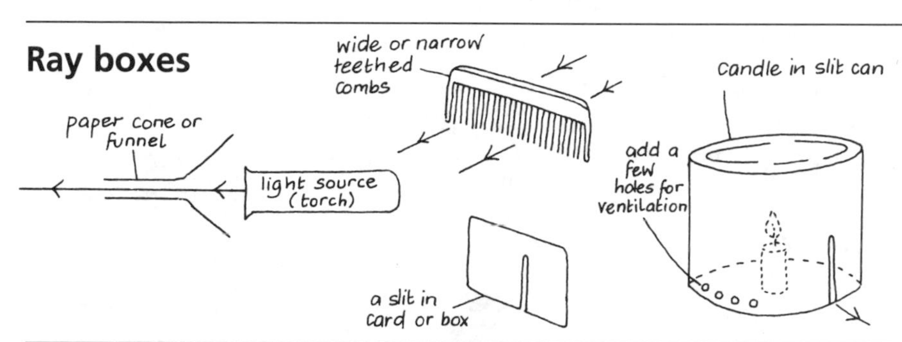

Many experiments with light require thin beams of light. Several methods of creating such beams are shown.

Investigate with students what else could be used.

Plastic and water lenses

You will need
- *2 sheets of bendy transparent plastic or perspex*
- *Plasticine or clay*
- *board*
- *water*

Bend the plastic sheets into either a convex or concave shape. Keep them in position by bedding them into the clay or Plasticine on a board. Seal the edges with Plasticine or clay. Fill the 'container' with water and it acts as a lens.

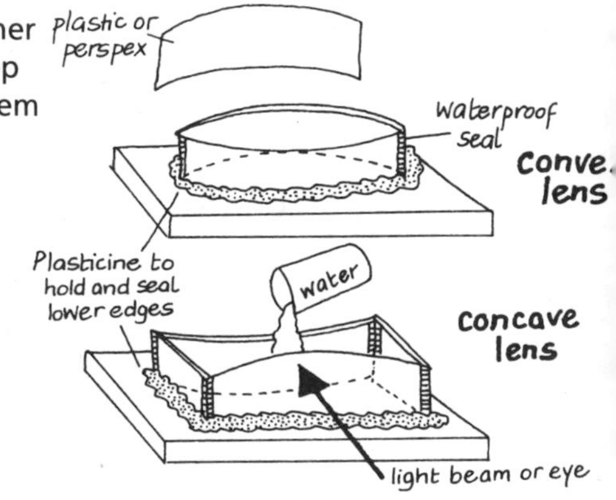

Reflection

Compare reflections in water, windows, tin cans and shiny surfaces.

Discuss why some reflective surfaces (mirrors) distort the reflection.

Refraction

Bending light

You will need
- *a pane of perspex or glass*
- *light beam*
- *sheet of white paper*

Lie a pane of glass or perspex on a table. Shine a light beam through one edge of the pane and note the way the beam is bent. The light is bent (refracted) at the interface between the glass and air. Note that both the beams in air are parallel.

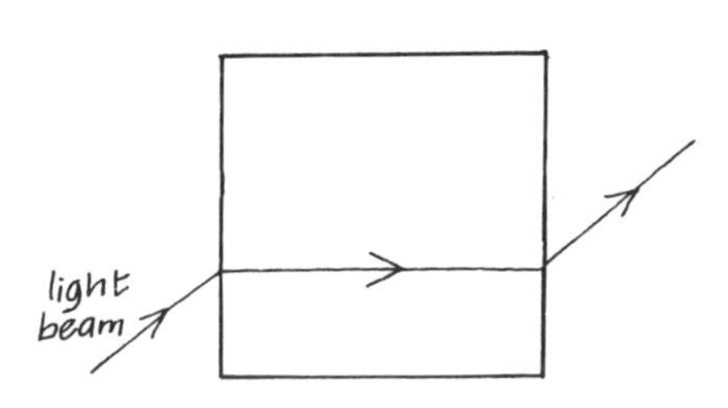

The rising coin

You will need
- *coin*
- *dish of water*

Place a coin in a dish or lid. Look across the edge of the lid so the coin is just not visible. Ask someone else to add water (so the eye does not change position). Note that the coin then becomes visible – it appears to have risen in the water.

Reflection and refraction

Light entering through shutters can be seen as fine rays. The dust particles in the air reflect some of the light that falls on them and they seem to shine.

The sun shines through tiny droplets of water in the sky. The light is reflected and refracted in every direction. Light seems to be all around us, not just a single large ray from the Sun.

Diffraction and interference

Put 2 fingers close together, leaving a very small gap. Bring the fingers very close to one eye, closing the other eye, and an interference pattern of vertical black lines can be seen in the gap between the fingers. The same effect can be achieved if 2 slits are cut in a piece of paper (cut or score the paper with a sharp edge, e.g. scissors or razor blade). The slits should be very close together and about the length of the index finger.

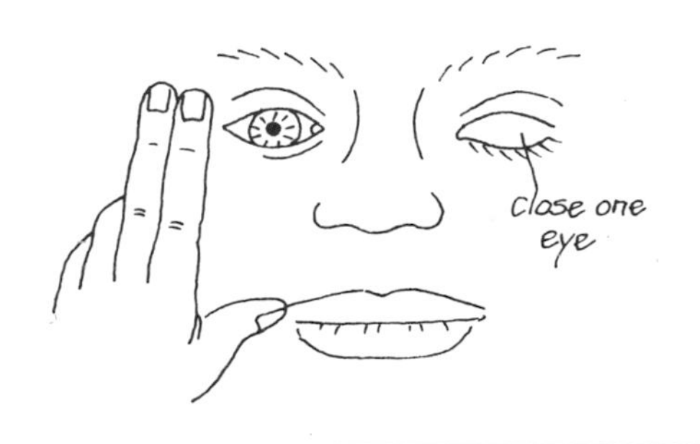

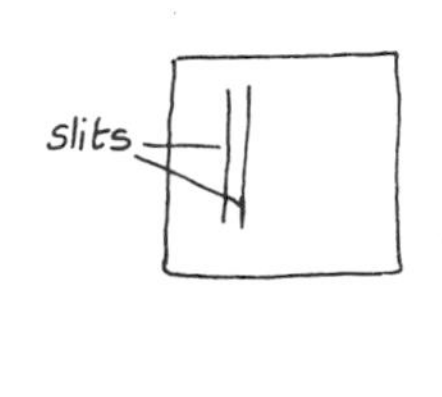

Pinhole cameras

You will need
- *cardboard tube, or tin, or lightproof box*
- *greaseproof paper*
- *dark, lightproof paper*
- *pin*

Make the cameras as shown. The pinhole should be small. The image is seen on the greaseproof paper. Ask students to note the image is upside down and is smaller than the original object being viewed.

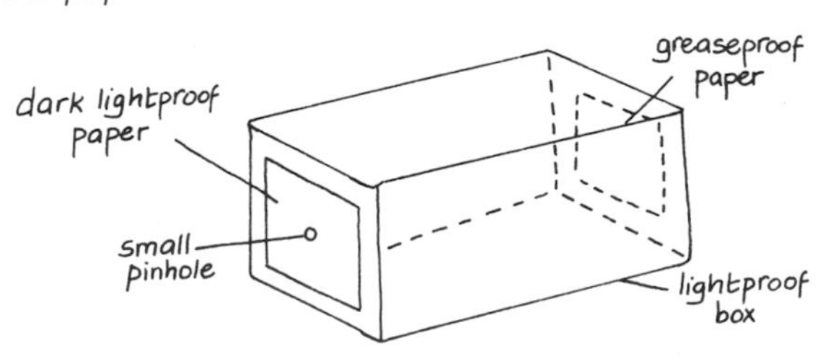

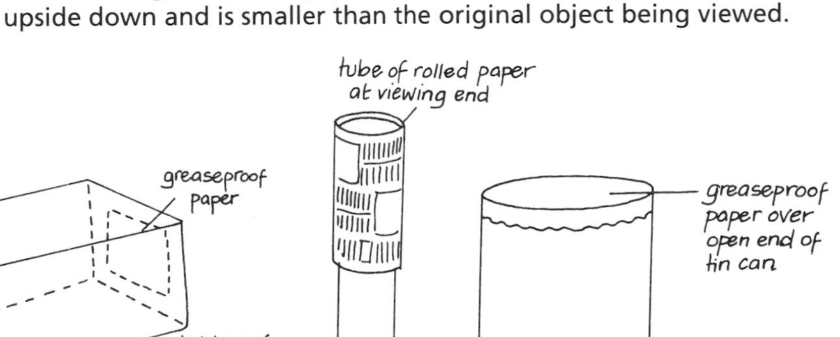

Periscope

You will need
- *2 mirrors*
- *cardboard tube or long box*
- *tape*
- *scissors*

Make the periscope as shown. Experiment with tubes and tin cans to make periscopes of different heights and shapes.

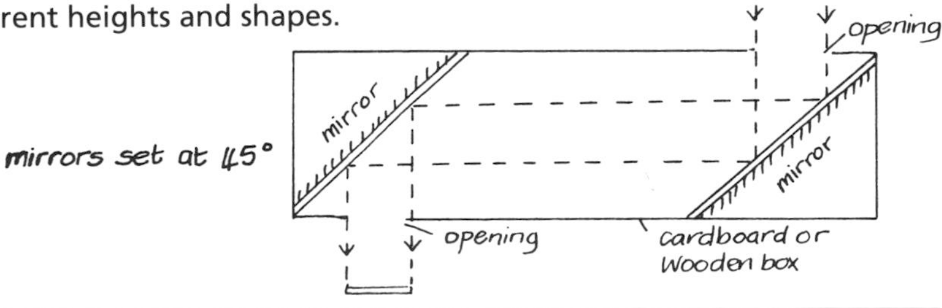

Kaleidoscope

You will need
- *3 mirrors*
- *rubber bands*
- *cardboard*
- *tracing or greaseproof paper*
- *small coloured objects*

Hold the mirrors together with the reflective sides pointing inwards. Wrap them in cardboard and fix with rubber bands. Seal one end with tracing paper or greaseproof paper. Put some pieces of grass or small coloured objects inside the tube and view.

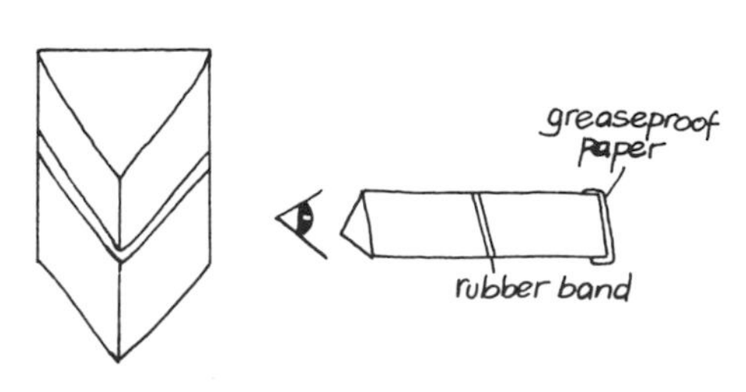

Colour

- **White light** (the light all around us) can be split into its **component** colours by, e.g. a **prism** or water drops.
- White light splits up because its component colours travel at slightly different speeds. Red light bends least, indigo most.
- Filters **absorb**, and so remove, light of particular colours or **wavelengths**.
- The colour of a substance depends on what wavelengths are absorbed by the substance. A red object, for example, looks red because all colours other than red are absorbed by the substance.
- Mixing coloured lights produces different effects to mixing the same colours as **pigments**, e.g. red and green light will produce yellow light.
- **Primary colours** are those which are needed to create all other colours. They are red, yellow and blue, for pigments, but red, green and blue for light.

Breaking up light

Water prism

You will need

- *3 small sheets of glass*
- *adhesive tape*
- *Plasticine*
- *Vaseline*

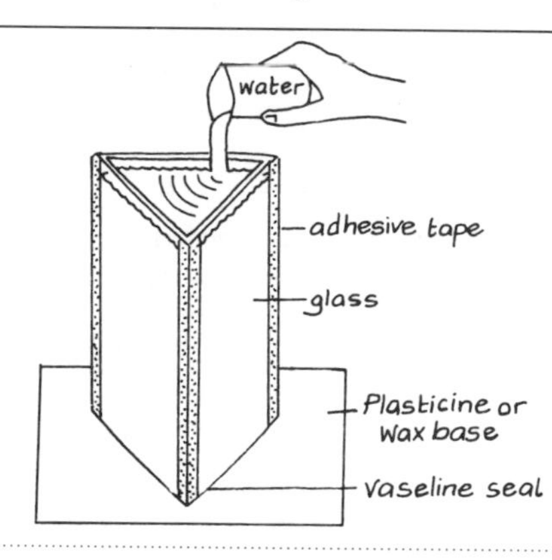

Stick the 3 pieces of glass together with tape. Use Vaseline along the joints to make them watertight. Push the prism into a base of Plasticine or candle wax so it is watertight. Fill the prism with water.

Investigate how the prism breaks up light into the colours of the rainbow by shining a beam of light through it. *(See page 98.)*

Soap bubbles

You will need

- *detergent or soap*
- *water*
- *sugar*
- *wire*

Bubbles refract the light as it passes through them. They split light into the component colours. Dip the wire shapes into a soap solution and blow steadily through the loop. Make the soap solution as follows. Mix 3–4 tablespoons of soap powder with 4 cups of hot water. Leave the mixture to stand for 3 days, then add a large spoonful of sugar.

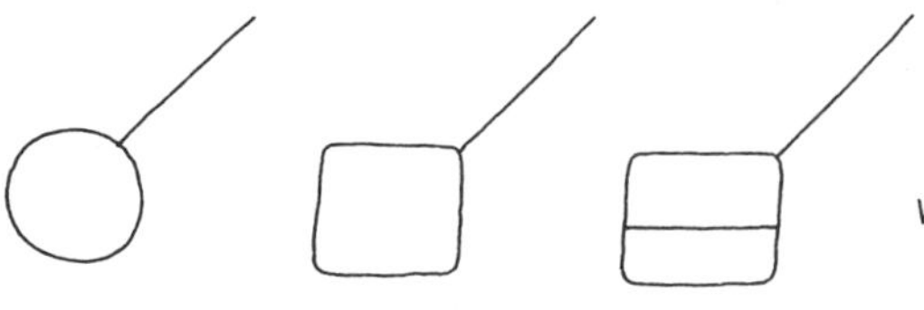

Investigate with students whether the shape of the wire affects the bubbles.

Mirrors and water

You will need

- *tray of water*
- *mirror*

Angle the mirror in the dish of water. Direct a beam of light (see page 98), or sunlight, through the water and onto the mirror. Project the light onto a piece of white card or a wall. The angle of the mirror and the water together act as a kind of prism.

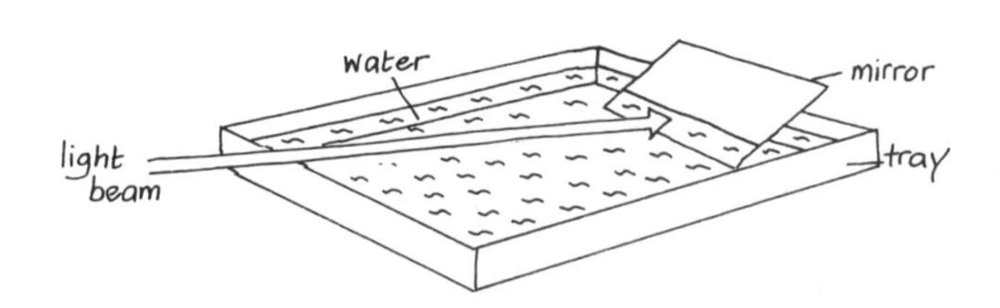

Place a convex lens in front of the white paper and direct the rainbow beam through it. Note that the rainbow pattern on the paper disappears because the 'rainbow' has been refracted again and the colours recombined.

Colour filters

You will need
- *coloured transparent materials, e.g. sweet wrappers*

Use a torch in a darkened room as a light source to look through filters of different colours. Note that filters absorb all colours but their own.

Shine coloured light on objects of different colours and note any apparent colour change in the objects.

Try mixing light from 2 different torches each with a different colour filter on it. Note the colour changes. Mixing red and green light causes some surprise!

Colour spinners

You will need
- *card*
- *sharpened stick as spindle*
- *primary colours of light (red, green and blue) in paint or crayon*

Key: R = red
O = orange
Y = yellow
G = green
B = blue
I = indigo
(each 51° for Rainbow Spin)

Make and colour the spinner as shown. Note the colours are not the pigment primary colours. Ensure the colours are as near to true primary colours as possible. When the spinner is spinning fast enough all the colours merge until the spinner appears to be white.

R O
O R

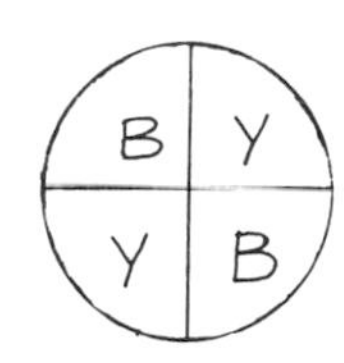

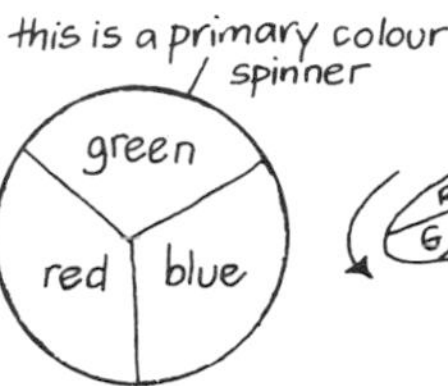

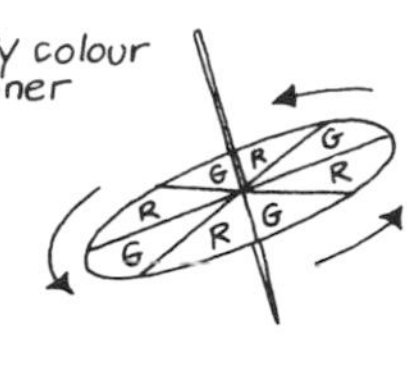

Experiment with spinners of different colours to see the effects, e.g. rainbow colours or just green and red.

Separating colours

The process of chromatography

You will need
- *coloured flower petals*
- *water or alcohol*
- *filter paper, blotting paper or white newspaper*
- *string*
- *dish of water*

Pound up the flower petals in a little water or alcohol. Dab a drop of the coloured liquid onto one end of the filter paper strip. Attach the strip to the string as shown with only the very edge dipped into the water in the dish. Alternatively dab the colour onto a filter paper cone as shown.

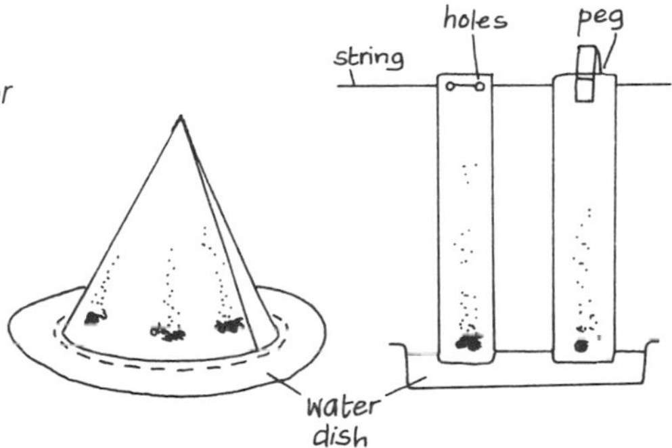

separating cone separating strip

Different coloured pigments do not travel at the same speed up the paper. This process separates the component pigments .

Experiment with pigments from foods, e.g. sweets, or inks.

Batik and tie dyeing

You will need
- *flowers, roots, fruits, etc.*
- *metal container*
- *molten candle wax*
- *cloth*
- *fine string*

Crush the flowers, roots or fruit and boil in water for at least 15 minutes. Strain the coloured liquid through a cloth into a bucket. The dye is now ready for use.

Experiment with different plants to find new colours. Some suggestions are:

- green – spinach or cassava leaves
- yellow – onion skins
- brown – tea, coffee, iodine
- blue – drops of iodine in warm flour solution.

For batik, the designs are drawn on cloth with molten wax. The cloth is then placed in the dye. Dye does not affect the waxed areas. After the dye has dried the wax is removed by ironing through paper.

In tie dyeing the cloth is pleated and then tied tightly with string. The dye does not penetrate the areas which are tied tightly.

Fluids and flying

- **Fluids**, e.g. water and air, push against (resist) anything moving through them. In the case of air this is called the **air resistance**.
- Objects are propelled through air and water by using their **resistance** as the **reaction** to the force of propulsion (**thrust**). (See page 88.)
- As air moves it causes changes in **air pressure**. A fast **flow** of air causes a drop in air pressure.
- **Streamlining** reduces turbulence or **drag** and so streamlined objects require less **energy** to move.

Air resistance

Parachutes

You will need

- *newspaper*
- *cotton material, e.g. handkerchief*
- *Plasticine*

A flat sheet of newspaper falls slowly, a crumpled up one faster. This is because the surface area for air resistance is greater in the sheet. Experiment with newspaper and note the more crumpled it is the more likely it is not to get blown off course.

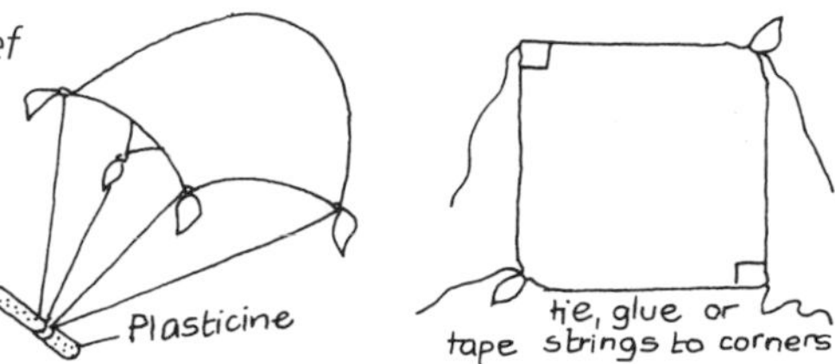

Make parachutes and investigate the effect of varying the Plasticine weight.

Make a small hole in the top of the parachute. Ask students if it flies better and why.

Natural seed parachutes

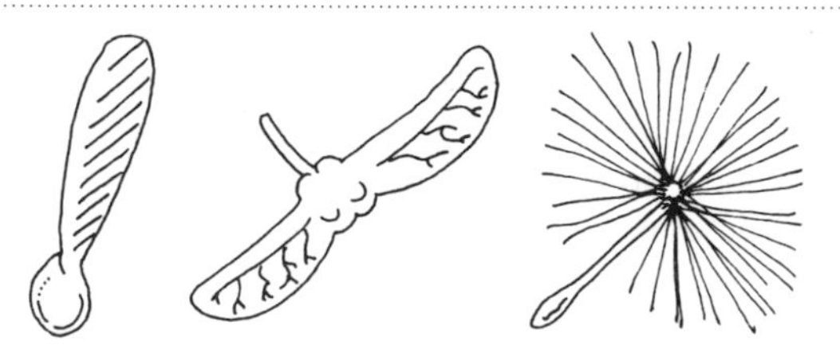

Find local seeds which have parachute devices.

Discuss with students why it is an advantage to a seed to have a parachute.

Helicopter

You will need

- *card*
- *scissors*

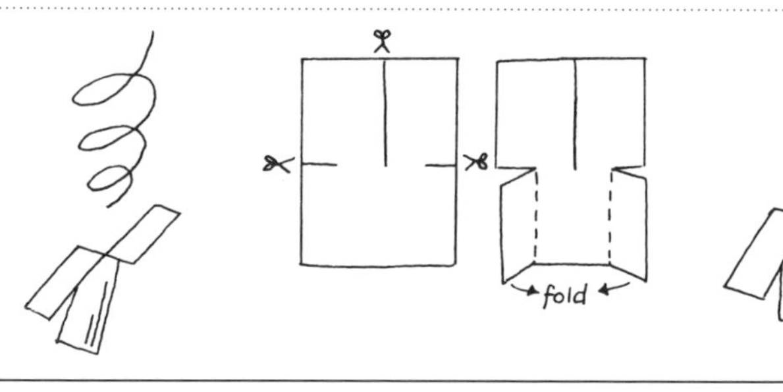

Make 'helicopters' from the card as shown. Experiment with different wing shapes and lengths.

Investigate which wing shape will 'fly' furthest from a testing place.

Forces in boats

Thrust

You will need

- *plastic container*
- *rubber tubing*
- *funnel*

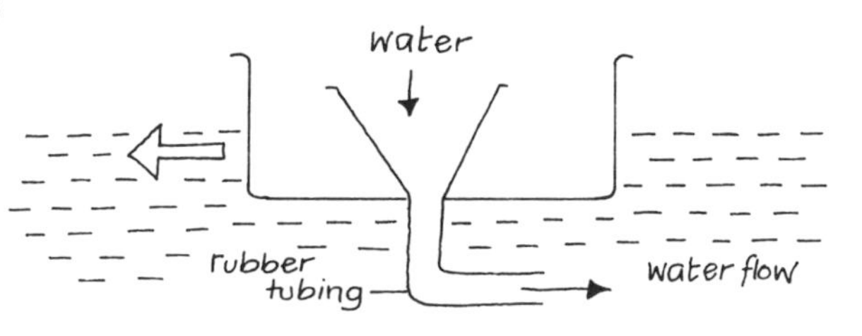

Make the 'glug-glug-tug' as shown. As water is poured into the funnel, the tug moves forward.

Experiment with different materials, e.g. polystyrene drinking cups, ballpoint casings, to produce the fastest tug.

Moving in small boats

Water is pushed backwards while the boat or canoe moves forwards.

As the stone is thrown, the boat moves in the opposite direction.

Discuss the forces at work in these 2 examples using ideas of reaction and action. (See page 88.)

Drag and streamlining

You will need
- *Plasticine*
- *tall transparent container*
- *water*

Make different shapes out of the Plasticine and test which sinks fastest. Students should find that shapes which are streamlined have less resistance, or drag, and sink more quickly.

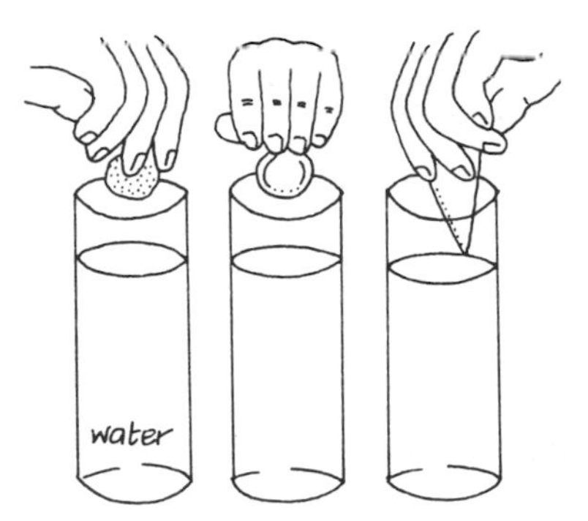

Study birds as they fly, swoop and dive and notice how they change their shape to make themselves more streamlined.

Examine fish and other water creatures and note which are most streamlined.

Ask students why is it an advantage to some water creatures to be streamlined.

Tricks with air flow

Tricks with paper

You will need
- *strips of paper*

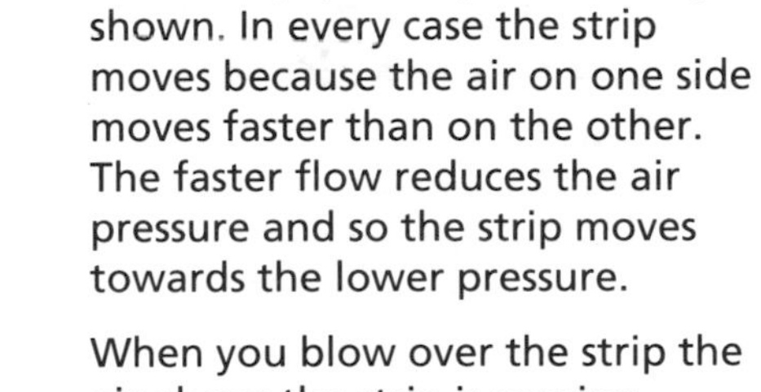

Blow the paper strips in the ways shown. In every case the strip moves because the air on one side moves faster than on the other. The faster flow reduces the air pressure and so the strip moves towards the lower pressure.

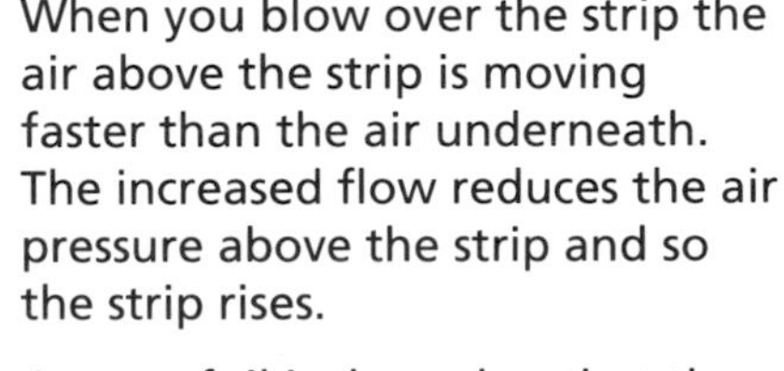

When you blow over the strip the air above the strip is moving faster than the air underneath. The increased flow reduces the air pressure above the strip and so the strip rises.

An aerofoil is shaped so that the air moving above the aerofoil has to travel further than the air underneath, i.e. it has to flow more quickly over the top surface.

The air between the sheets is flowing, that outside is not. So, the air pressure between the sheets is reduced. The same effect will pull 2 light fruits or ping pong balls together.

Tricks with ping pong balls

You will need
- *ping pong ball or light fruit*
- *funnel*

Blowing down the funnel causes the air to rush past the ball which reduces the air pressure. The air pressure outside the funnel then pushes the ball up the funnel.

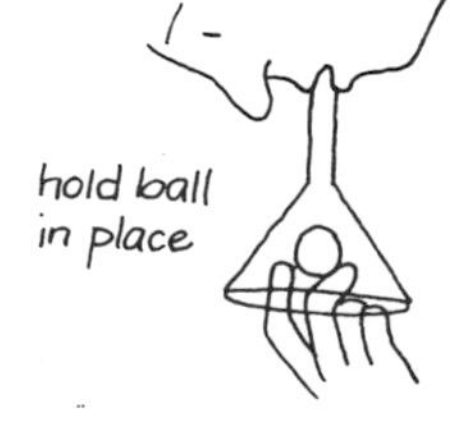

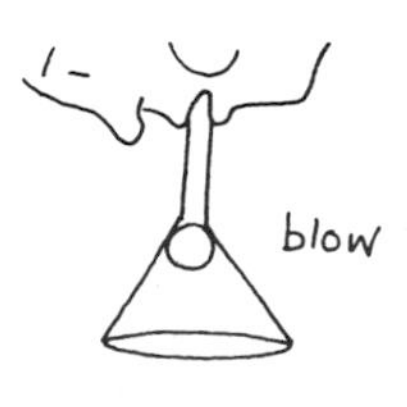

Making up chemicals; preparing gases

Safety: **Wear safety goggles when preparing chemicals.**

Note: Always use clean water when preparing solutions. Use distilled water or rain water.

Refer to page 126 for local sources of chemicals.

Limewater

- Mix lime in the water. Add 10 g of lime (CaO or $Ca(OH)_2$) to 1 litre of water.
- Allow it to settle.
- Decant off the clear liquid. This is limewater.
- Store limewater in a tightly stoppered bottle or jar to prevent absorption of CO_2 from the air.

Sodium hydroxide solution

- Slowly dissolve 330 g of solid sodium hydroxide in 500 cm^3 water and dilute to 1 litre.

Benedict's solution

- Make 2 solutions A and B then mix the 2 together to make Benedict's solution.

Solution A

- Dissolve 100 g anhydrous sodium carbonate and 173 g sodium citrate in 800 cm^3 water.
- You may need to heat the mixture.
- Filter if necessary – a precipitate may form.

Solution B

- Dissolve 17.3 g copper(II) sulphate in 100 cm^3 of water.
- Dilute to 150 cm^3.

Mixing solutions

- Mix together all solution A and all solution B, then dilute to make up to one litre.

Fehling's solution

- Make 2 solutions A and B which are then mixed.

Solution A (copper(II) sulphate solution)

- Dissolve 34.7 g of copper sulphate in 500 cm^3 water.

Solution B (alkaline tartrate solution)

- Dissolve 173 g of potassium sodium tartrate (Rochelle salts) and 50 g of sodium hydroxide in water.
- Dilute when cold to 500 cm^3.

Mixing solutions

- Mix equal volumes of solutions A and B when needed.

Biuret's solution

- Make a copper(II) sulphate solution by dissolving 37.4 g copper(II) sulphate in 500 cm^3 water.
- Make sodium hydroxide solution as described above.
- Mix equal volumes of the solutions when needed.

Preparing gases

Carbon dioxide

A test for carbon dioxide is that it will put out a lighted splint or match and will turn limewater milky.

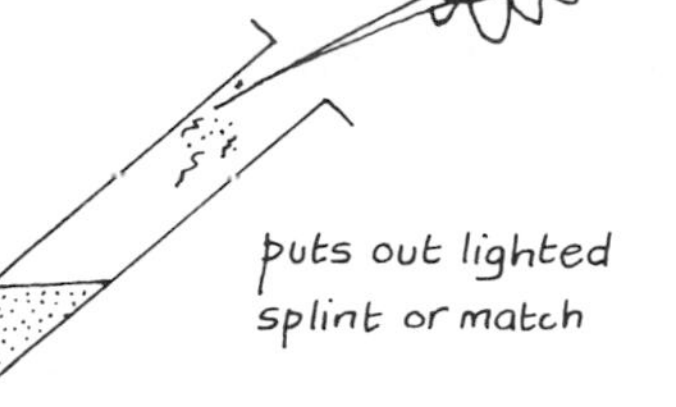

1 Mix together vinegar and ashes or vinegar and soda (bicarbonate of soda, baking soda or baking powder).

2 Add water to Andrew's Liver Salts.

3 Mix yeast and sugar solution and keep them warm. The carbon dioxide will fill the balloon, the alcohol simply dissolves in the water.

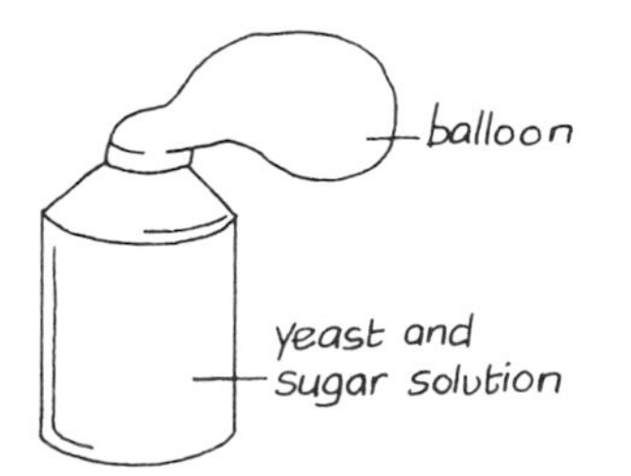

Chlorine

Chlorine is a poisonous gas.

- Dissolve potassium permanganate in concentrated hydrochloric acid.
- Use 5 cm^3 concentrated acid to 1 spatula of potassium manganate(VII).

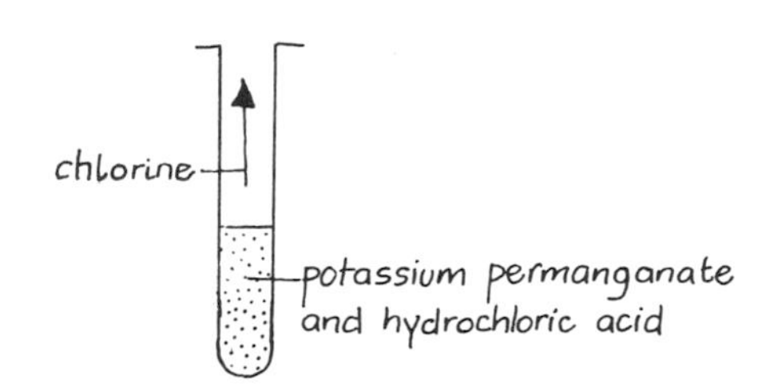

***Safety:* Concentrated hydrochloric acid burns. Safety goggles should be worn when using it.**

- **Skin contact – wash with plenty of water.**
- **Eyes – wash the eye immediately with large quantities of water.**

Hydrogen

- Dissolve zinc (from dry cells) in hydrochloric acid. Hydrogen gas is produced.

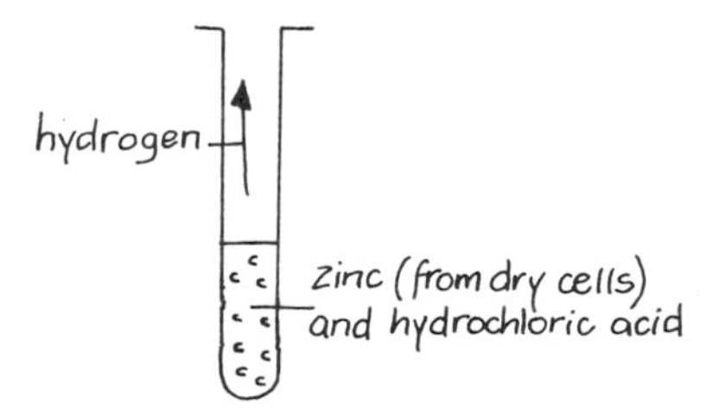

***Safety:* Hydrochloric acid burns. Safety goggles should be worn when using it.**

- **Skin contact – wash with plenty of water.**
- **Eyes – wash the eye immediately with large quantites of water.**

Oxygen

Safety goggles should be worn.

A test for oxygen is that it will relight a glowing splint or match.

- Mix manganese dioxide with hydrogen peroxide. Oxygen is produced.
- Alternatively heat potassium manganate(VII).

relights a glowing splint
or match

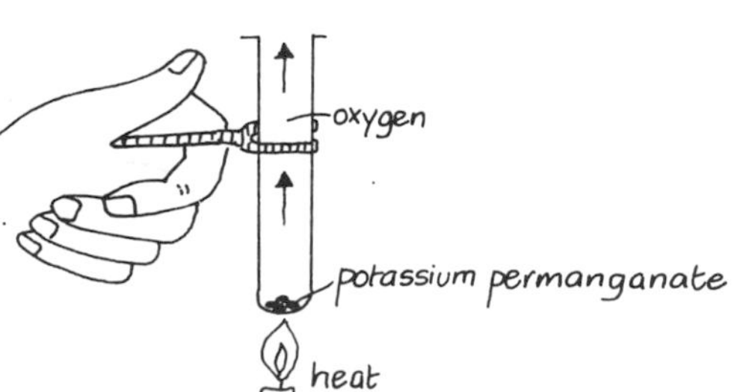

Laboratory equipment

The following pages show how teachers can make substitute equipment, when standard laboratory equipment is unavailable. Home-made equipment may not always be quite as effective or as safe as commercially produced, standard equipment and teachers should be particularly careful about safety when such equipment is used. Take extra care if used by students rather than in a demonstration by the teacher. Ensure students wear safety goggles when they use home-made equipment, just in case it fails.

Using what is available

- How many experiments can be carried out with everyday items?

Multi-purpose bottle

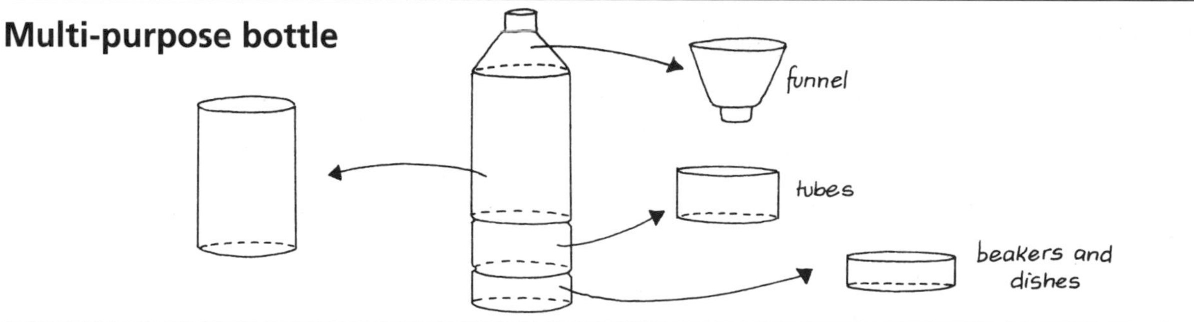

Test tube holders

- Folded paper strips, strips of card or clothes pegs all work well.
- Form a wire test tube holder by shaping a piece of wire round a piece of wood the same size (diameter) as your test tube.

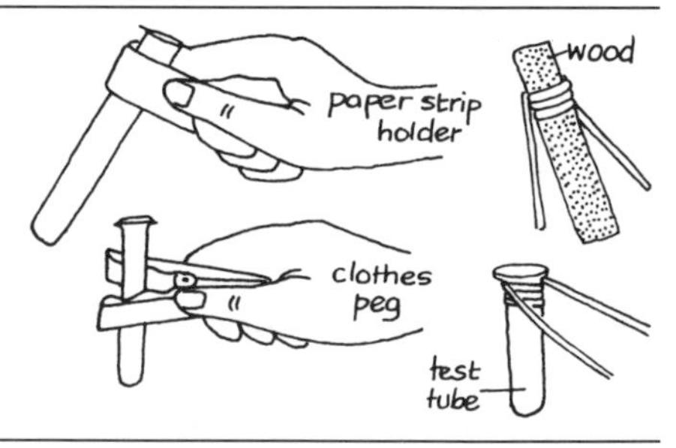

Funnels

If a paper funnel is covered with foil its life is extended.

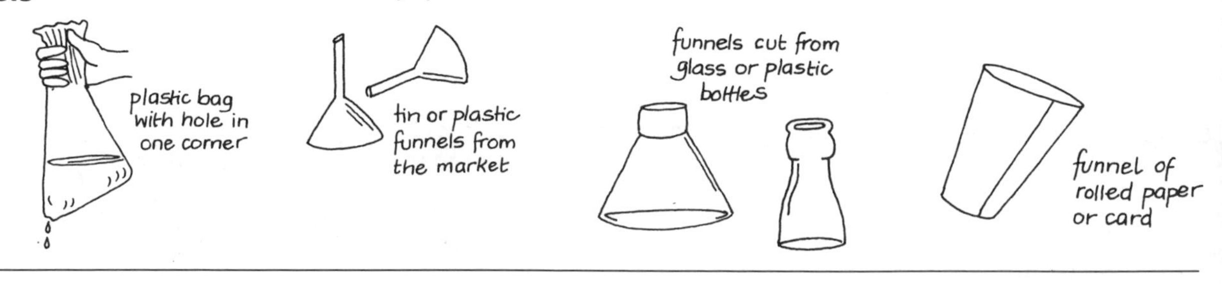

Filter paper

Test different types of paper – not printed papers because the colours may run. (Use cement bags, copy paper, etc.)

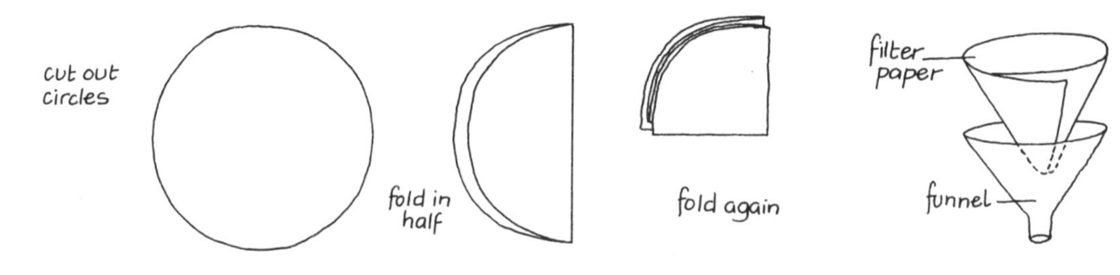

Beakers

Safety: **Such containers may shatter if they get hot. If the container has to be heated, standard laboratory equipment must be used.**

Cups, glasses, jars and bottles are useful for experiments at room temperature.

Light bulbs as multi-purpose containers

- Fused electric light bulbs can be used as containers.
- They should never be heated as they may explode.

Opening and emptying the bulb

- Always wrap the bulb in a thick cloth for safety when working on it.
- Rub the base over a rough surface (cement or stone) to weaken the seal.
- Alternatively, use a pair of pincers to remove the seal.
- A nail can be used to open and clean out remains of the seal.
- The filament and rest of contents come out easily.
- File the opening smooth if needed.

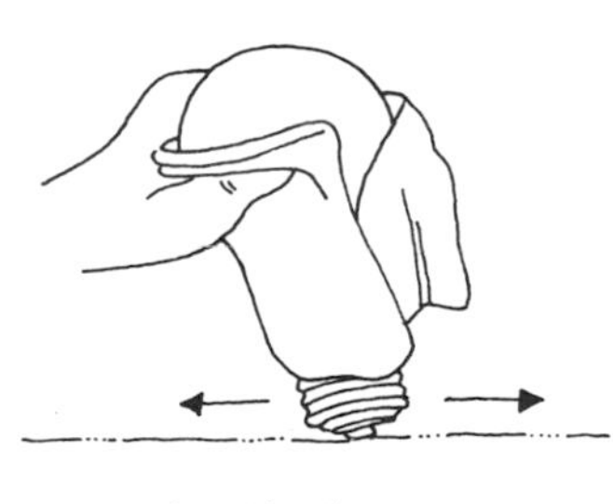

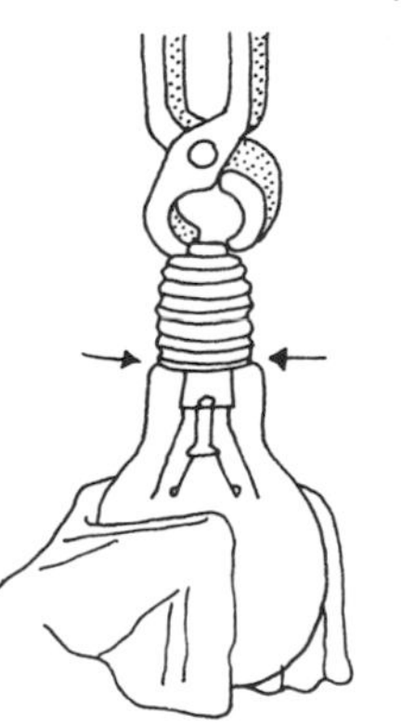

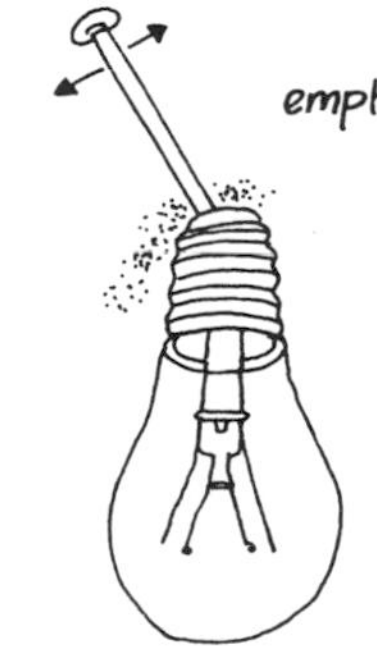

Using the container

- A bored bung/cork with an attached rubber tube may be added.
- **Do not use if contents will get hot. Never heat in a flame.**

Tubes

Glass tube substitute

- Everyday items can be adapted for use as tubes.

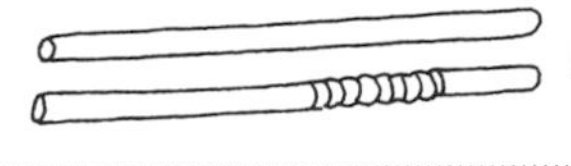

U-tubes

- Use bent plastic pipe or hosepipe.
- Ball point casing will bend if heated gently.

Laboratory equipment *continued*

Blowpipe

***Safety:* Use safety goggles when using a blowpipe.**

- This will allow you to produce a higher temperature flame or a directed flame.
- A fine, heat proof tube is used to blow the flame from a burner.
- The metal refills found inside some ballpoint pens work well.

Combustion spoon

- This is used to heat small quantities of material in a flame.
- Ensure the bottletop is clean and the inner plastic seal removed.
- Wrap the wire around the bottletop to form a cradle for it.

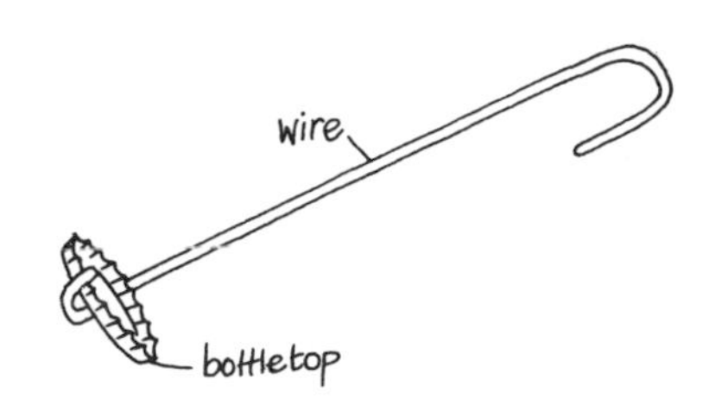

Crucible

- This is used to heat material to a high temperature.
- Place the material in one spoon and then wire 2 spoons together.

Gas generator

You will need
- *a small bottle or similar container*
- *a small piece of a ballpoint inner tube*
- *wider flexible tube*

- Make sure the tube does not become blocked.
- Never use concentrated acids or hydroxides in the apparatus.
- Use only small quantities of substances in the apparatus.

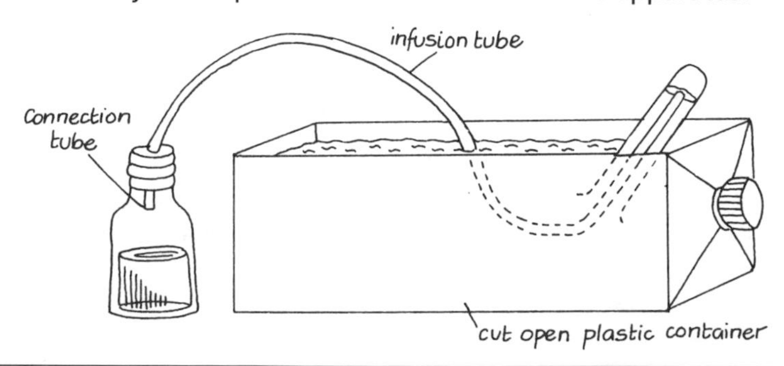

Bell jar and trough

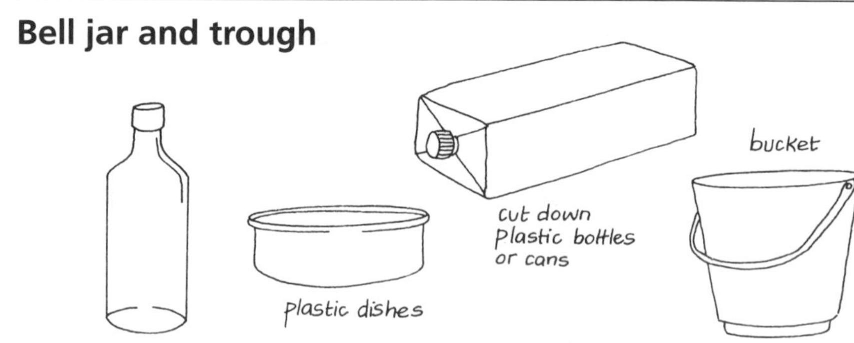

- Bell jars can be made from large plastic bottles if the bottom is removed.
- Note plastic will only be suitable for some experiments.
- Troughs can be made from any vessel which is large enough.
- Large plastic containers are useful.

Stirrers

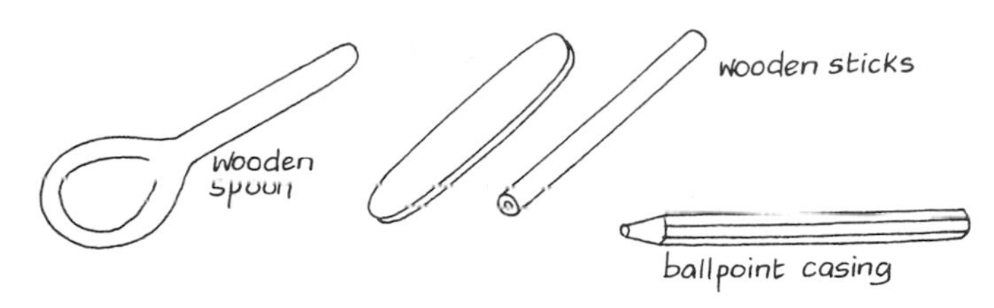

- These should be made from materials which do not carry heat so they can be used for hot materials.
- Select materials carefully when heated substances are involved.

Spatulas

- Ordinary plastic spoons are useful, but there are many alternatives.

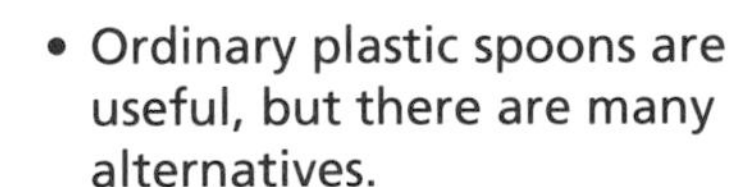
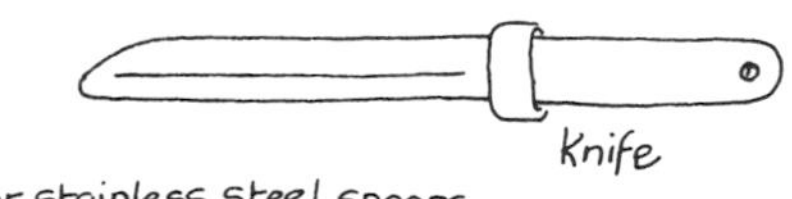
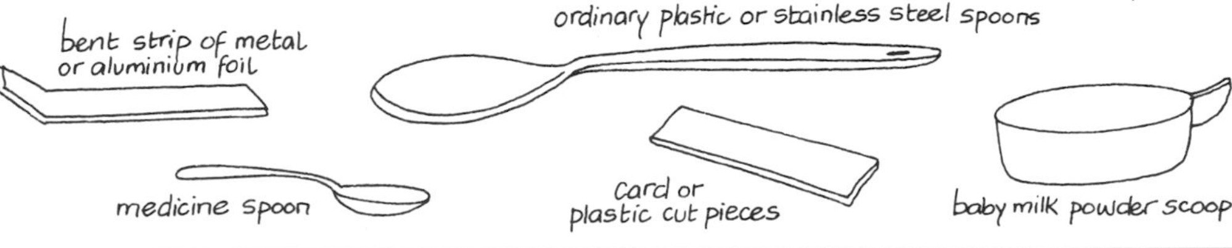

Stoppers and corks

- These can be cut and shaped from soft woods, maize cobs, cork or pieces of rubber.
- Sand or file to shape.
- Explore the market place – rubber tops from a variety of bottles may be available.
- Can you use the cork or bung in its original bottle for your experiment?

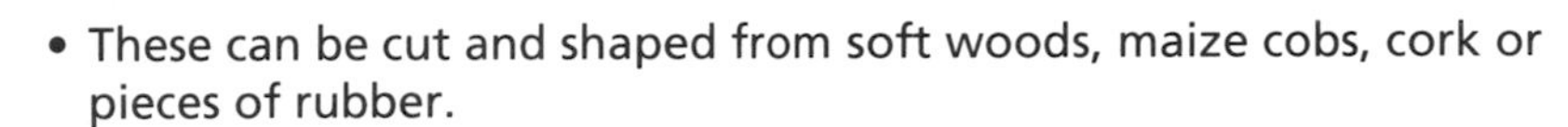

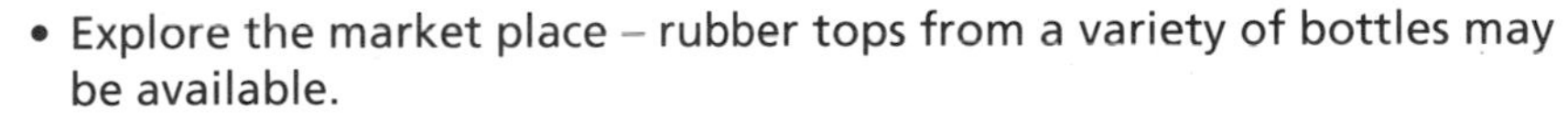

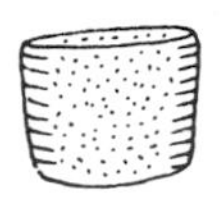

Wire gauze

- A tin lid with holes in it will work.
- Try wire mosquito netting.

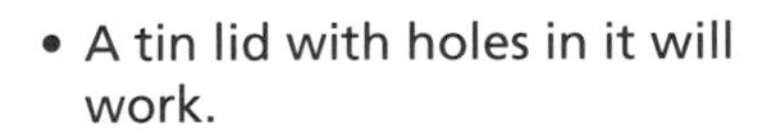
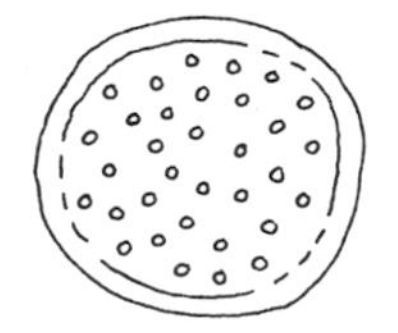

Pestle and mortar

- A flat grinding stone works very well.
- You can crush items using the back of a spoon.
- Try using different sized spoons – one as a pestle the other as a mortar, and crushing material between them.

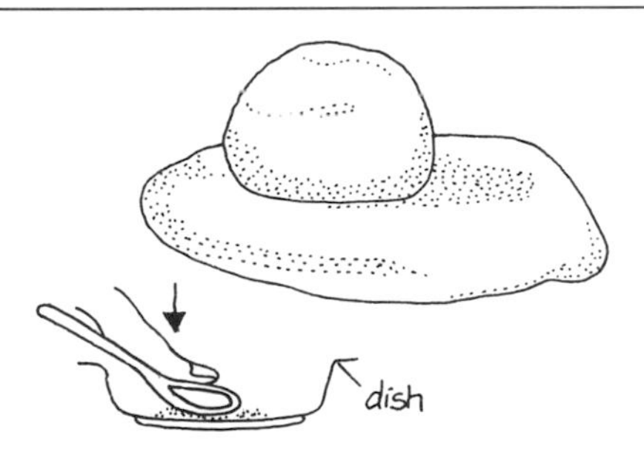

Test tube brush

- Put a little sand in the test tube as an abrasive. Move the sand around with a stick. Or make a brush as shown.

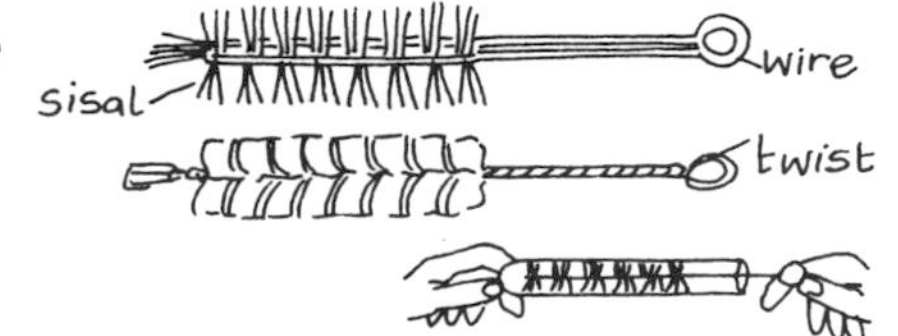

Test tube rack

- Experiment with your own ideas.

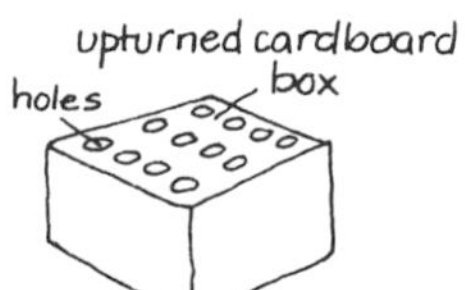

Laboratory equipment *continued*

Droppers

- Droppers from medicine bottles.
- Ballpoint casing with a bulb made from rubber tube.
- Could you adapt a fountain pen?

Tongs and forceps

- Metal packing strip can be bent and then cut with tin snips.

Tripods

- Thin wire can be used to join together pieces of galvanised wire.
- The bottom may be removed from a metal can and the sides cut with tin snips to leave 3 legs.
- 3 legs are more stable than 4.
- Experiment with your own ideas.

Supports for apparatus

Funnel holder

- A large loop of wire makes a good base for the support.

paper strip

wire

Wire supports

- Use a piece of wood the same diameter as your test tubes to prepare supports.

Wood

Pipettes

- Glass rods, straws and ballpoint pen casings can all be used.
- The larger bore, transparent tube of a ballpoint casing can be calibrated with 'scratch' marks.

Cover air hole

Handles

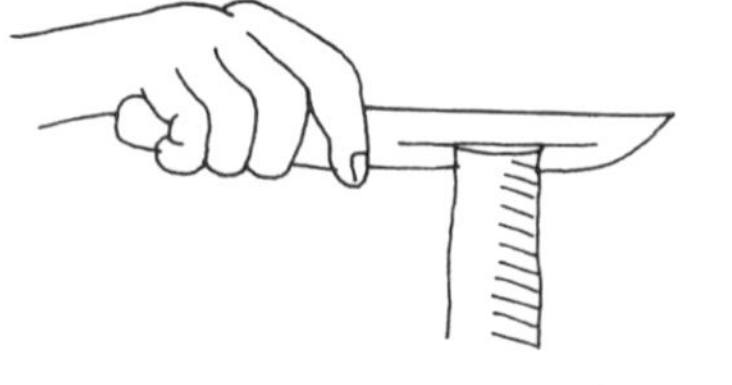

biro casing – a needle mount

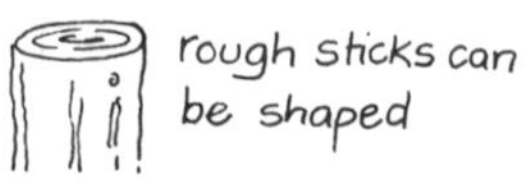

pin or needle handle – an eraser (rubber)

Measuring tools

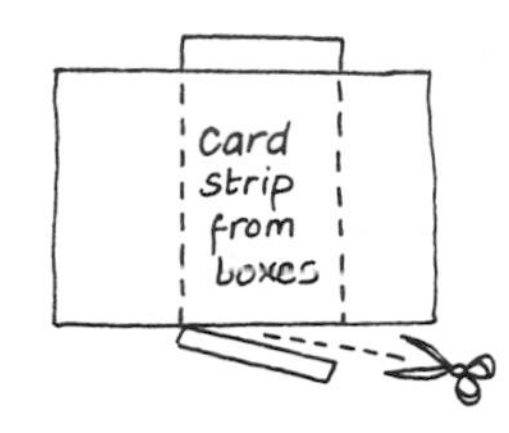

- Students can make these themselves.
- Metre rules can be made from various items.

metre rule from cloth strip

split bamboo or palm branch
(try guinea corn stalks/other grasses)

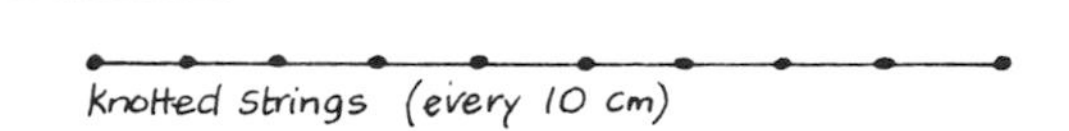

Measuring cylinders

Safety: These tubes cannot be used with organic solvents.

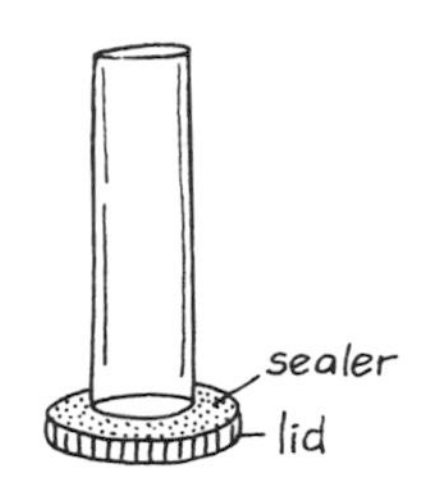

- Cut an empty fluorescent light tube to length.
- Seal one end with a lid containing wax, tar, cement or adhesives (see page 118).
- After drying, graduate using a standard measuring device or a syringe.
- Sterilised syringes are suitable for measuring small quantities.
- Plastic syringes may dissolve in some substances.

***Sterilising:* Boil hard for 15 minutes in boiling water.**

Calibrating cylinders

- Jars, tumblers and cut down bottles (see page 113) can all be used as measuring devices.
- Grind graduation lines on glass bottles with a glass cutter or triangular file.
- Scratch marks on plastic vessels or use spirit markers or tape.
- Marker inks can be made by mixing 2 parts borax with 10 of water, then mixing this with 6 parts of alcohol and 1 of shellac or varnish.

Overflow bottles and cans

- The simplest device is to put one container inside another, larger one. Fill the smaller one with water, put in the solid and measure the water which overflows.
- An overflow can is made by cutting 2 slits at the top of the can and bending the strip forward to form a spout.
- In the overflow bottle, the water is level with the top of the biro casing until the object is put in. It is useful, therefore, to be able to move the biro casing up and down.

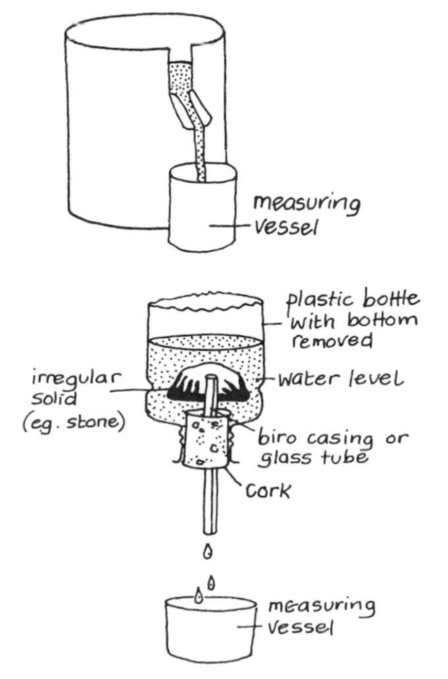

Laboratory equipment *continued*

Hydrometer

- Mark the scale on paper or plastic strip and put it into the straw or hollow grass.
- Drop a few stones or small tacks into the bottom to make the hydrometer float vertically when it is placed in a container of liquid.
- Coat the whole straw with candle wax and seal the end.
- Try making a hydrometer from a pencil.

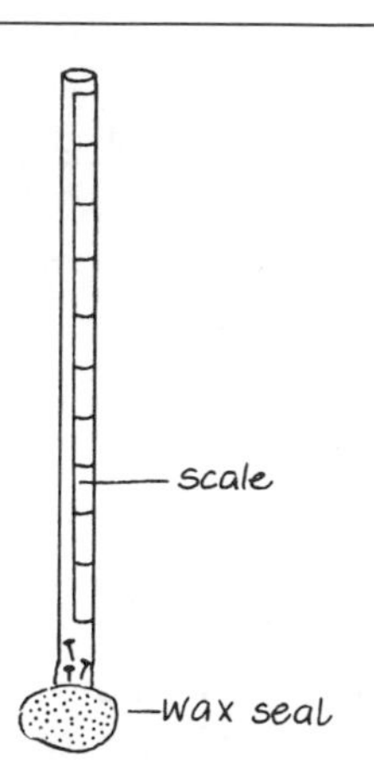

Beehive shelf

- The bottom of a varnished (to prevent rusting) tin can, or a plastic bottle, can be used as the base.
- Cut out a large hole in the top and a slit in the side to take tubing.

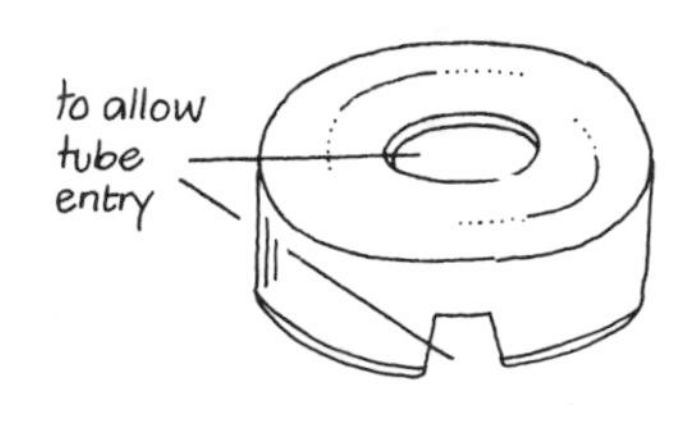

Tube connectors

- Cut pieces of metal tube from a metal ballpoint pen refill.
- Heat them in a flame then clean the tubes with wire.
- Push a section inside the tubing.

Making mirrors

- Make a reflective surface by painting one side of glass black.

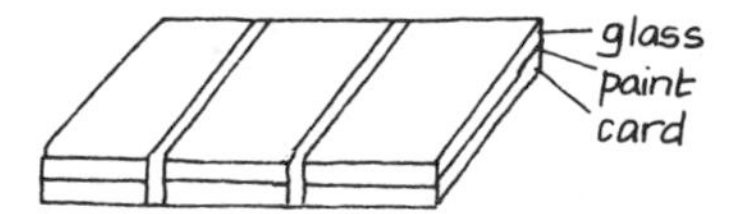

- Place aluminium foil between glass and card.
- Experiment with other foils from sweet and cigarette wrappings.
- Louvre glass may be suitable for making large mirrors.

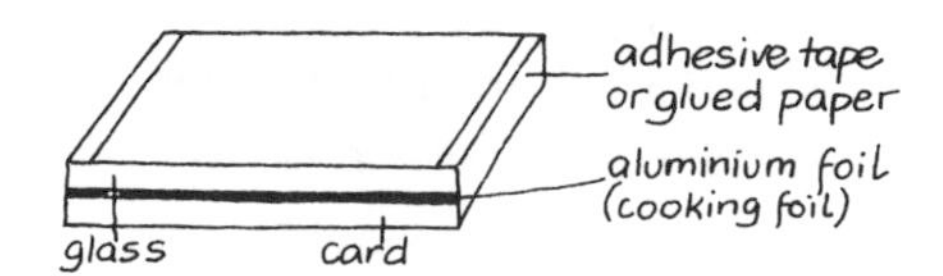

- The inside of a smooth, shiny tin can acts as a mirror, but tape the cutting edges for safety.

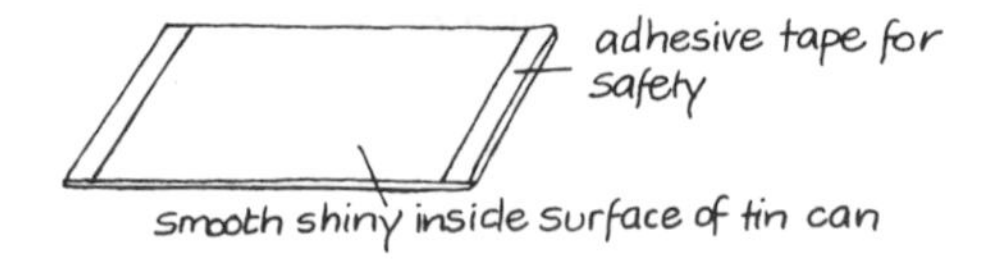

Glass cutting

***Safety:* Small pieces of glass are difficult to see and dangerous. Use safety goggles when breaking glass.**

Using a glass cutter

Work with gloves or a cloth for safety.

Safety goggles should be worn.

- The wheel of a glass cutter scratches a line on glass. This weakens the glass along the line so it is easier to crack the glass accurately.
- Dipping the cutter wheel in turpentine helps.
- Use a straight edge to score a line on glass. Mark the score with a pen so you can see it.

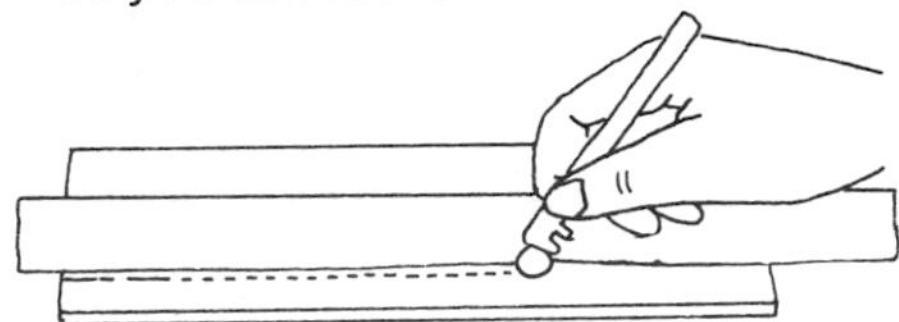

- Only a single, firm scratch is needed. Avoid scratching several times as this may give a jagged break.
- Break glass over a pencil or match, or by placing the score just over the edge of a table.
- Firm, gentle taps along the score mark should make the glass break cleanly.

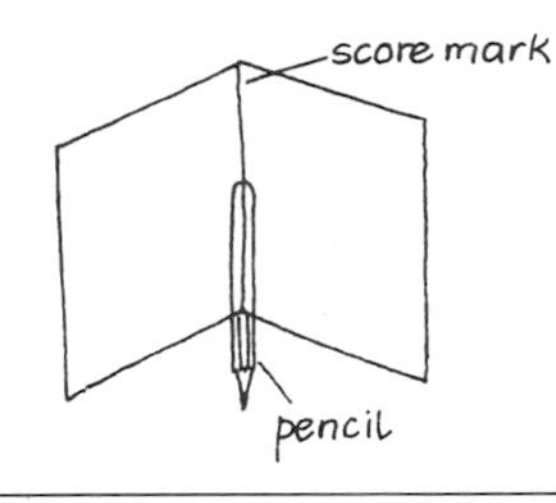

Cutting glass tube

- Use a triangular file to scratch and weaken your tube at the point you wish the tube to break. One good scratch/score is suitable for most tubes of diameter up to 25 mm.
- The scratch does not need to go all the way round the tube.
- Break the tube over a pencil or match as with glass sheet.
- Alternatively, grip the tube with both hands, thumbs touching immediately behind the scratch (scratch points away from you). Snap carefully as you would a small twig.

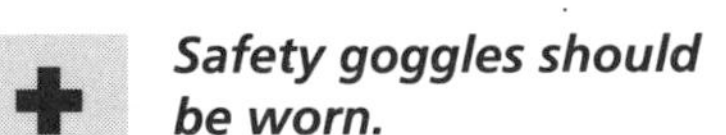

Safety goggles should be worn.

Cool strips cracking

- Make a mark with a file right around the bottle.
- Soak paper strips in water.
- Place a strip of soaked paper on either side of the scratch.
- Heat the scratch with a flame while rotating the bottle.
- Drop water on the scratch if necessary.

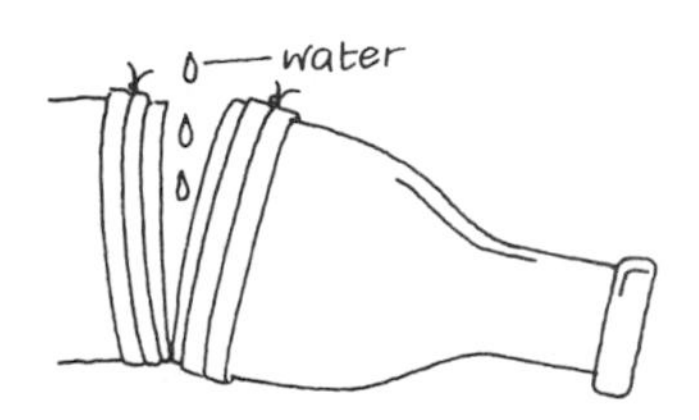

Safety goggles should be worn.

Magnifying and microscopes

Empty light bulbs

- Varying the amount of water in the bulb alters the magnification.

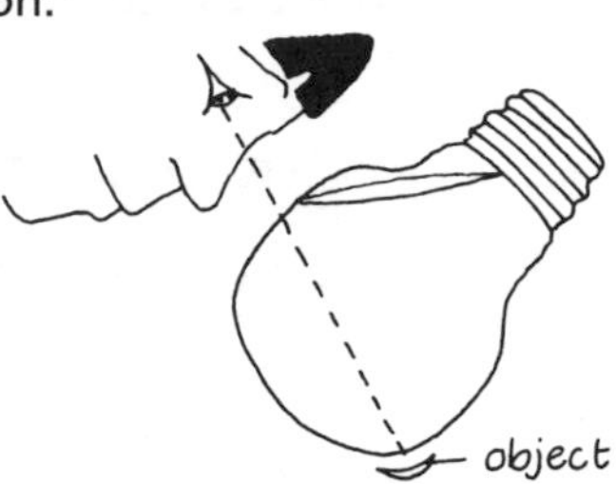

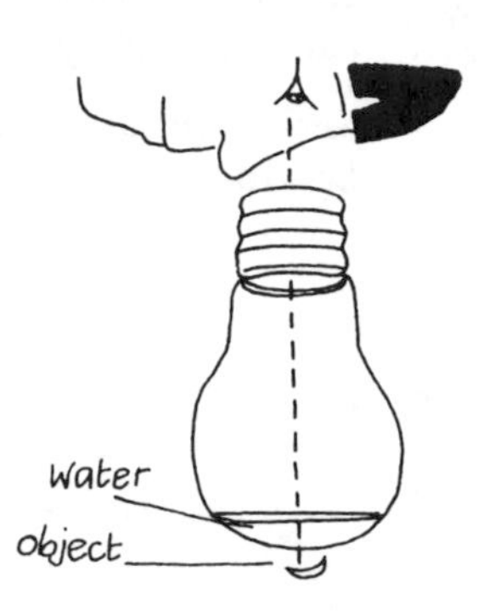

Water drop

- Make a hole in a piece of card or foil which will 'hold' a drop of water in it.
- When using drops to magnify, try to get the most efficient shape. The larger the drop the better.
- Experiment with different materials to find which gives the best drop.
- The wire ring from a broken light bulb filament will also hold a drop of water well.

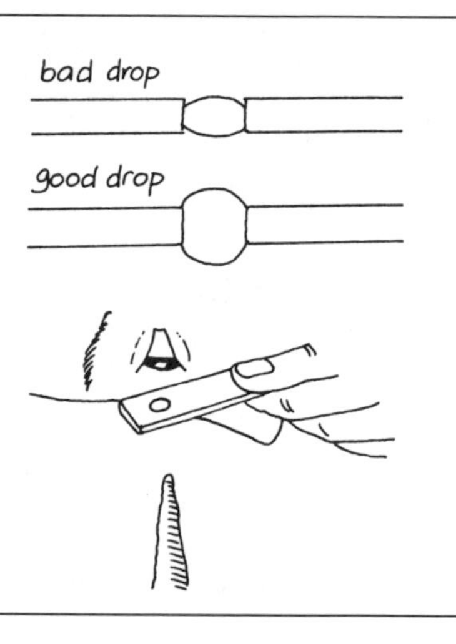

Curved glass

- Putting water into curved glass alters the magnification.
- Experiment with different depths of water.

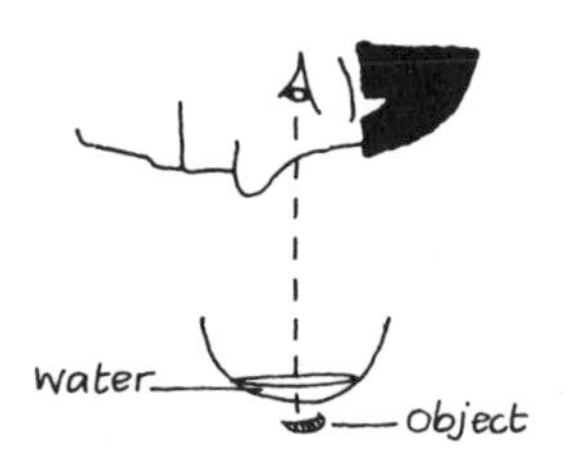

Clear-container magnifiers

- Any of these containers filled with water will make good magnifiers.
- A clear marble magnifies. Try marbles of different sizes.

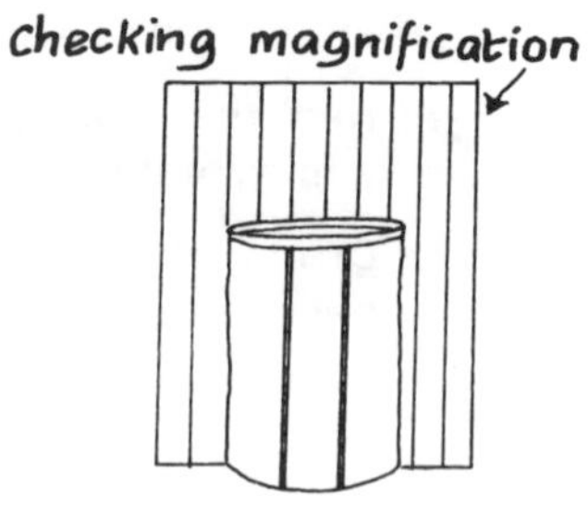

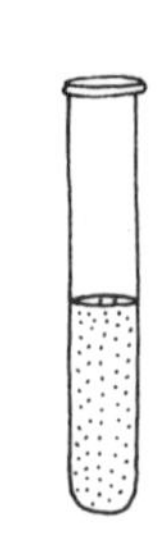

Old lenses

Spectacles

- Experiment with lenses of old and new spectacles – even broken spectacles have their uses!

Pencil torch bulb lens

Safety: *Beware of tiny bits of glass.*

- Remove the lens of the bulb with a triangular file or, as carefully as possible, with scissors or a knife.
- Mount the lens in a hole in a piece of cardboard.
- The lens will be held in position if the bottom of the hole is smaller than the top, or it can be held in place by tape.

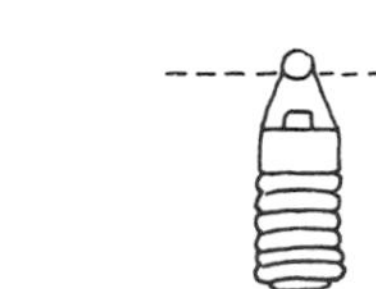

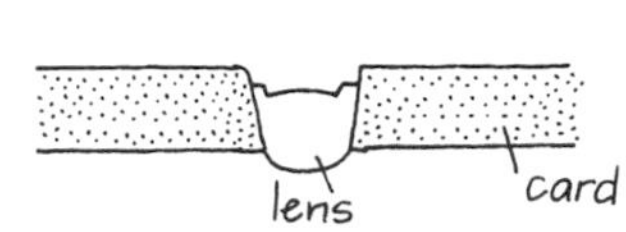

Simple compound microscope

- Using 2 lenses together allows much greater magnification.
- Use a hand lens to make a water drop into a more powerful magnifier.
- Try using a hand lens with a lens from a torch bulb to make another simple compound microscope.

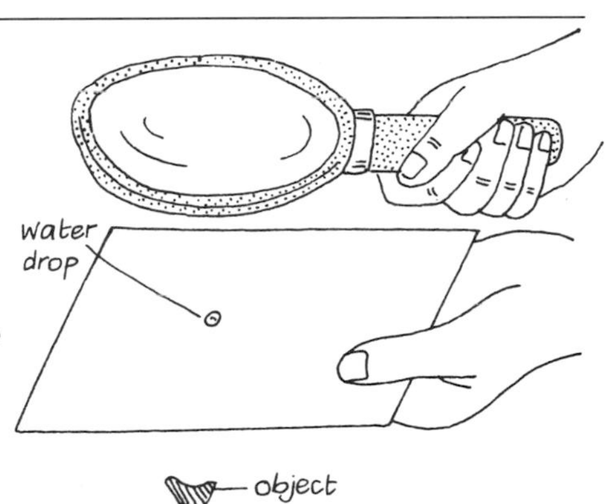

Card bridge microscope

Safety: *Ensure the mirror does not reflect direct sunlight into the eye.*

- Place a water drop in the card 'bridge'.
- Place this on a sheet of glass as shown.
- Place the object you are looking at on the glass. This arrangement is most suitable for thin items, e.g. sections of leaves.
- Experiment with the angle of the mirror so that light shines up through the specimen.
- Use this arrangement with a handlens to produce a compound microscope.

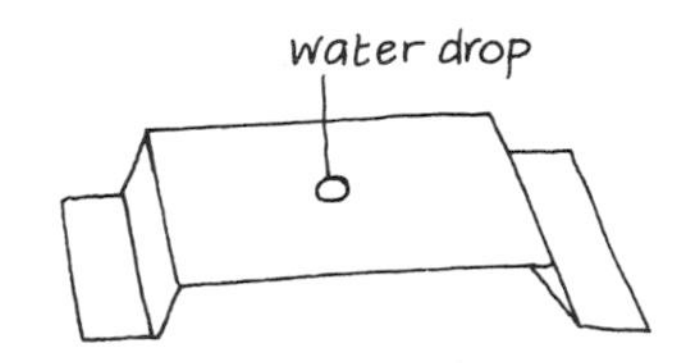

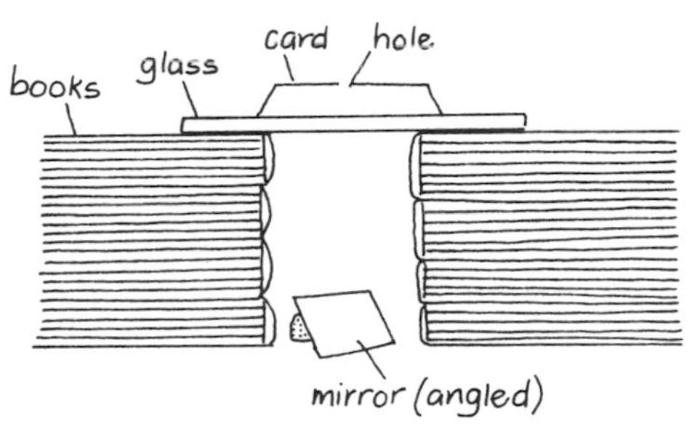

Slides and cover slips

- Small specimens will be clearer if placed on a slide and covered by a coverslip.
- Slides can be made from glass or transparent plastic.
- Coverslips may be round or square and can be made from thin, stiff, transparent plastic like that used in display packaging.
- The glass from some torches makes a good slide and can be put back into the torch after use.

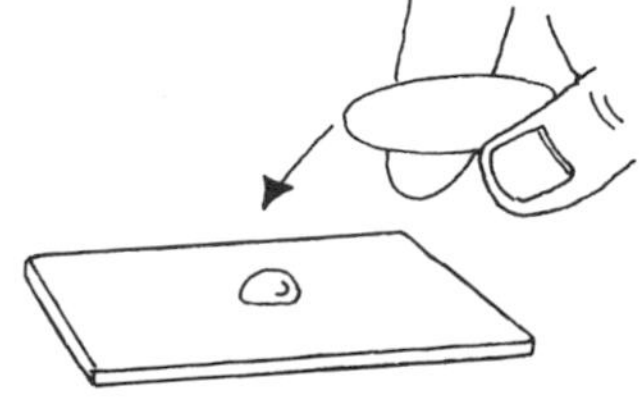

Burners

Safety warning: **All burners are potentially dangerous.**

- **Never use petrol in any burners.**
- **Only refill burners when they are cold.**
- **Do not make large spirit burners.**
- **Use small quantities of fuel.**
- **Wear safety goggles.**

Candle burner

- Fix 3 or 4 candles onto a dish, lid or piece of wood.
- Vary the heat by using 1, 2, 3 or 4 candles for efficiency.

Fat or wax burner

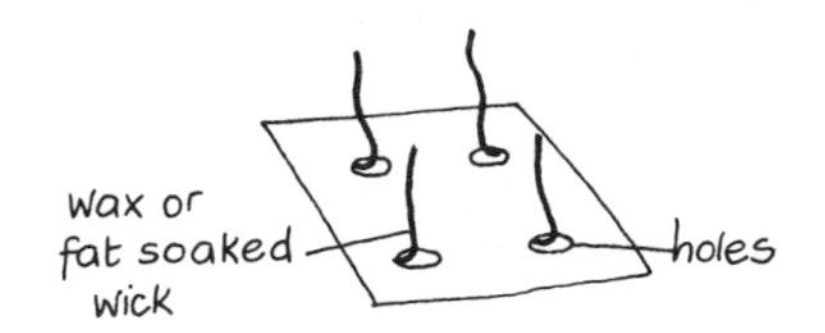

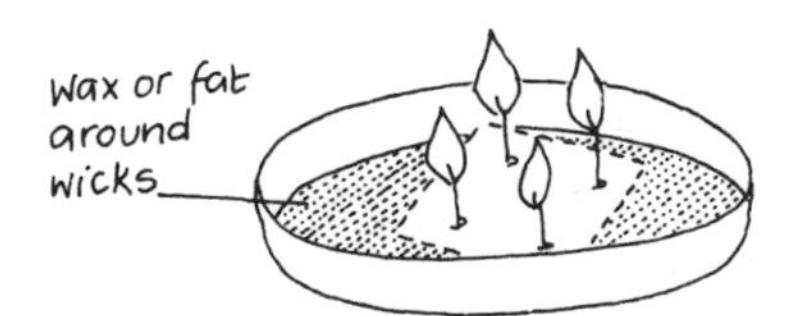

- Soak the wicks in fat or wax first.
- Push the wicks through holes made in card or tin.
- Float the card or tin, with the wicks in place, in a dish of molten wax or fat.

Simple spirit burner

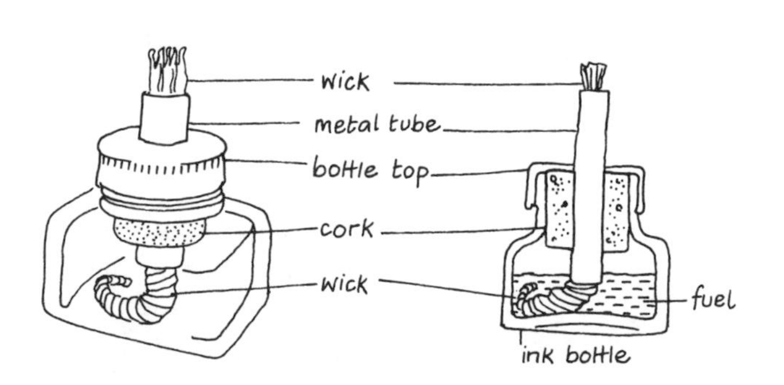

Valve spirit burner

- Push the wick through the bike valve which itself fits tightly into the lid of the tin.
- What burners are available locally?
- Could you adapt the local burner?

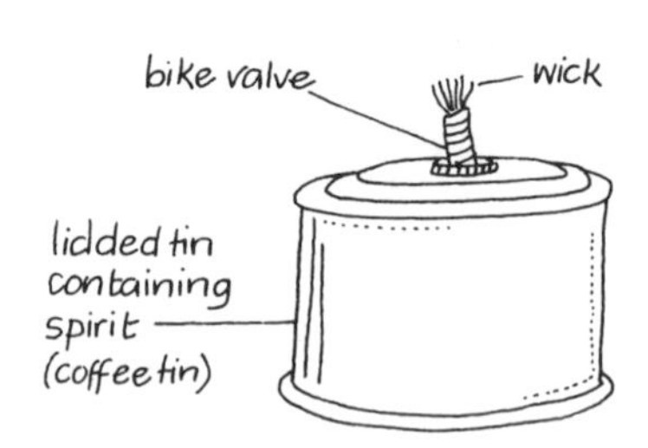

Mini charcoal burners

- The two-can version is easier to make, but you will need to experiment with different sizes and types of tin cans.
- Ensure the container does not fit too tightly into the top can. Wire supports may help.
- The temperature of the charcoal depends on the amount of air flowing through it and the amount of charcoal used.

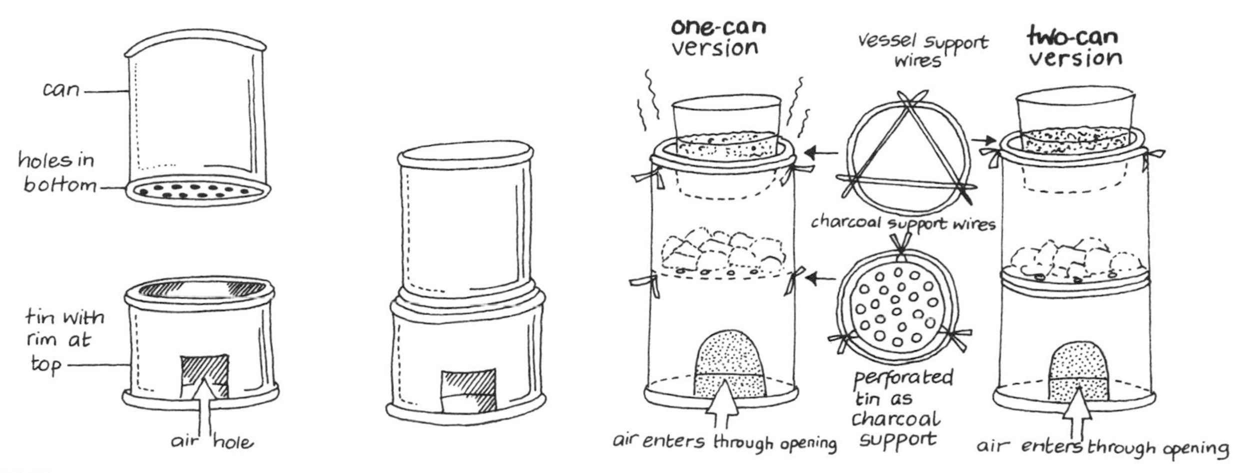

Simple sootless kerosene burners

The Pakistan model

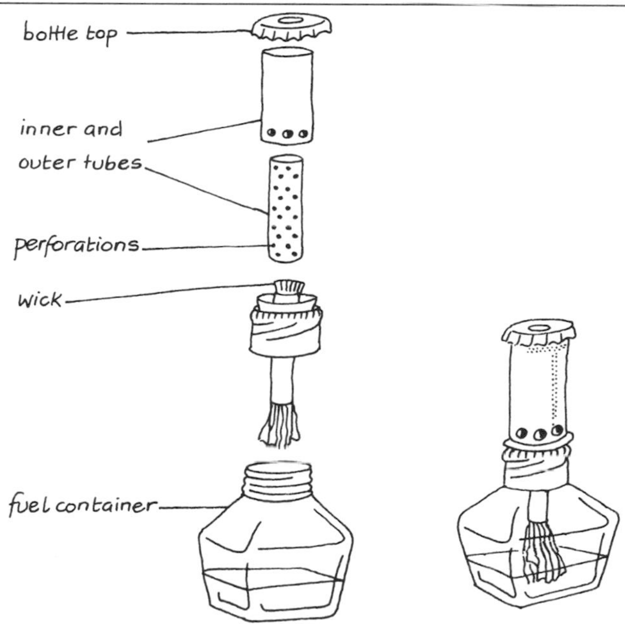

Temperature of different types of heating source

Type	*Average flame temp. °C*	*High point temp. °C*
candle	650	700
kerosene burner	650	800
alcohol burner	650	800
matches	600	650
Bunsen burner	1400	1500
candle and blowpipe	800	
alcohol burner with blowpipe	1000	

Joins and adhesives

Joining things

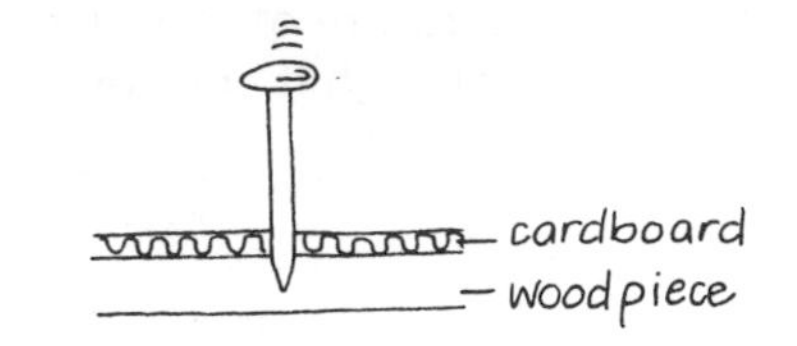

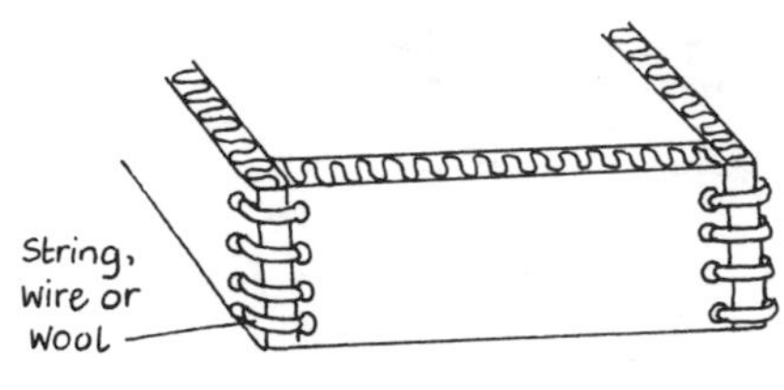

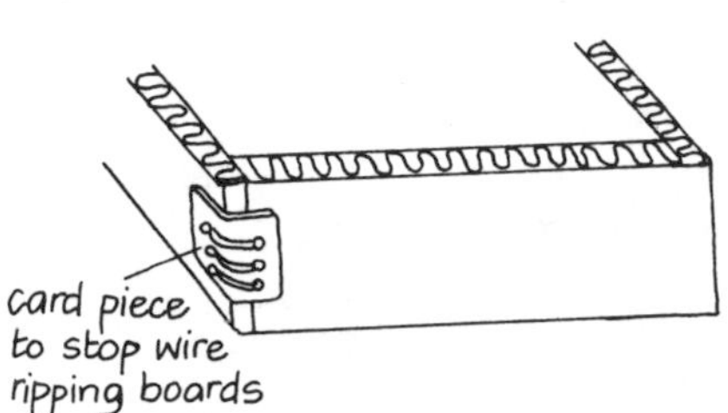

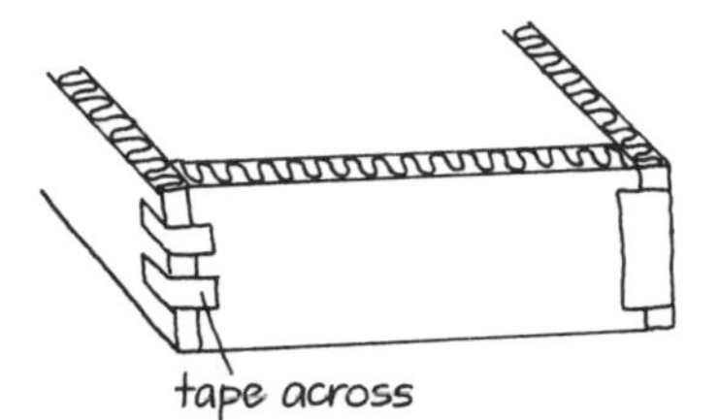

- Make holes with a nail, resting the surface on a piece of wood.
- Could you use the hinged corners of boxes? Think before you cut!
- A strip of adhesive tape, pasted paper or cloth stuck to the inside and outside of the corner makes a secure join.
- String, wire or wool can be used to join corners.
- Find your own method.

Cardboard joints

- In the hole and peg technique, a long peg can be stuck down at the back for extra strength.
- Never cut the slot deeper than half way for an interlocking slot joint.

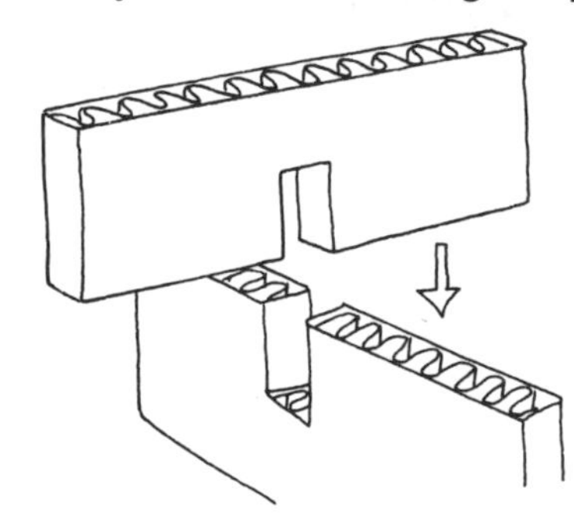

Flour paste

- Sift flour to remove lumps. Maize, wheat and cassava flours are all suitable.
- Mix the flour with water a little at a time to avoid lumps. It should be the consistency of thin cream.
- Cook the mixture gently until it thickens. Keep stirring to ensure the paste remains smooth and of even texture.
- Allow the paste to cool.
- Add insecticide to the paste if needed.
- Store in a clearly labelled container with a good lid, preferably in a cool place.
- Cold method paste is made by simply stirring sifted flour into water.

Polystyrene cement

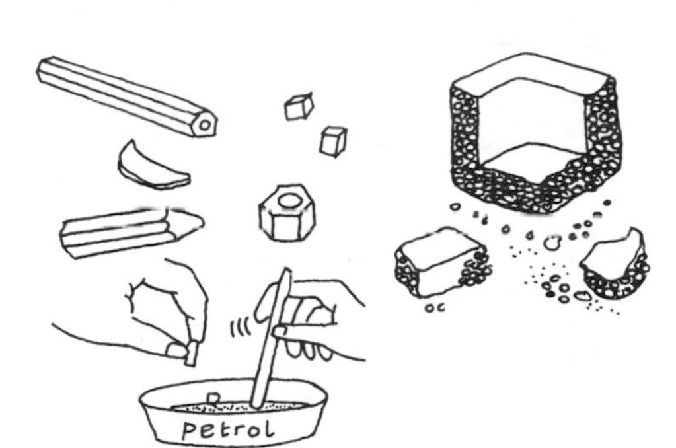

- Dissolve polystyrene ballpoint casings, or styrofoam packing material in a little petrol.
- Natural rubber dissolved in petrol will make 'rubber solution' – another glue.

Danger: **Fire hazard – flammable.**

Rice paste

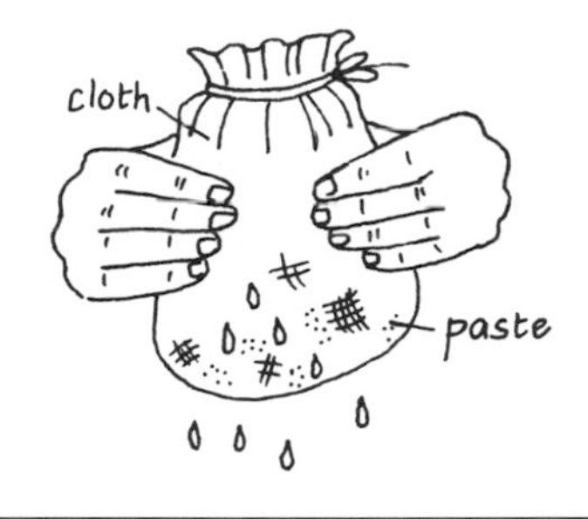

- Cook 1 part rice (white rice is better than brown for paste) with 3 parts water.
- Strain off excess water.
- Rice thickens as it cools.
- Squeeze the wet sticky rice through open-weave cloth.
- Squeezing breaks down the rice grains so they form a paste. The paste may be thinned down with water if necessary.
- The water which is squeezed out is also an effective light adhesive.

Casein glue

- Mix milk with vinegar or lemon juice. Add just enough vinegar or lemon juice to curdle the milk. The amounts will vary according to the type of milk used.
- Heat while stirring continuously. Soft lumps will form.
- Strain out the lumps using a cloth.
- Add a teaspoon of sodium hydrogencarbonate (bicarbonate of soda) to the lumps and mix with a little water to produce casein glue.

Fish or animal glues

These glues are made traditionally by boiling a soup of water, skins bones and sinews. However, they dissolve in water, so may not be effective in a climate which is both hot and humid. They work best in dry climates.

- These glues may be found in markets as brown blocks (a bit like toffee).

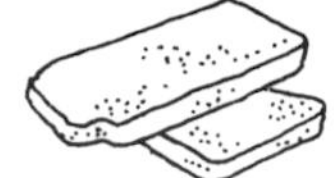

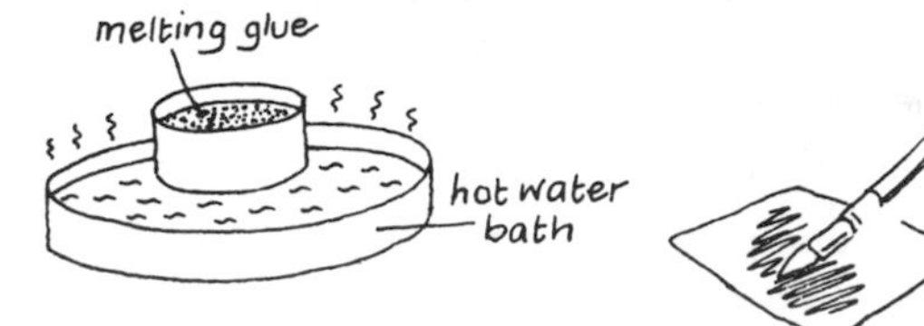

- Melt the block gently using a water bath.
- Use the glue when it is hot and runny.
- This type of glue is used for sticking wood in furniture-making and for attaching labels to food cans.

Modelling materials

Papier mâché

- Soak pieces of paper or card in water for half a day.
- Mash, grind, stir or pound the mix to a smooth fine pulp.
- Squeeze or press out excess water.
- Mix in a little flour paste (see page 118) and work the material into a sticky modelling consistency.

Papier mâché layering

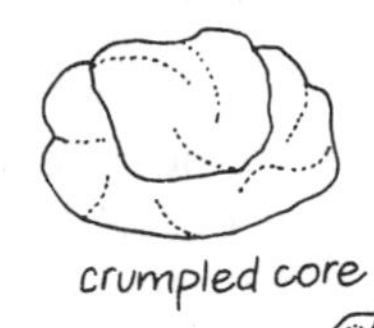

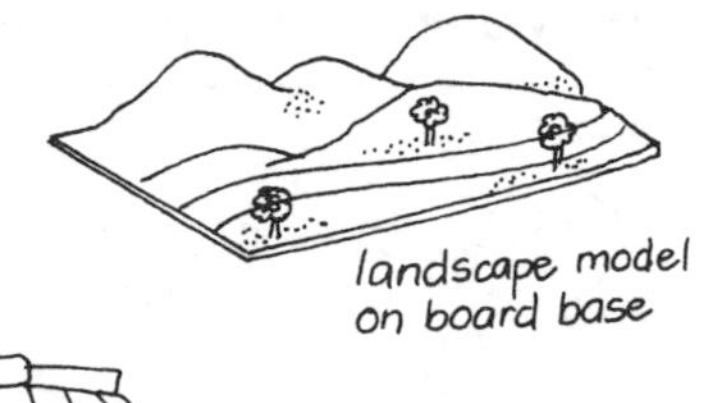

pasted strips

- Soak small pieces, or narrow strips, of newspaper in paste.
- Use crumpled newspaper as a core or skeleton for the model.
- Build up the model in layers of strips and pieces.
- After drying, sandpaper smooth and paint or varnish.

Modelling clay

- Dig out or collect your clay. Seek local advice on where to find suitable deposits.
- Add water and stir to a creamy consistency.
- Filter through cloth or a sieve.
- Allow the filtered material to settle.
- Decant excess water.
- Dry the filtered material on newspaper until it becomes a powder.
- Mix in glycerine to give a plastic texture.
- Knead well and add Vaseline to soften if necessary.
- Adding paste (see page 118) to the clay helps stop it cracking as it dries.

Modelling wax

- Melt 10 parts wax. (Ordinary wax candles work well.)
- Slowly add 2 parts turpentine, or turpentine substitute, to the melted wax. This makes it malleable.
- Add one part edible oil to the mixture.
- Add colouring if available.

vegetable oil

Modelling dough

- Mix together the following
  2 cups flour
  1 cup salt
  2 teaspoons cream of tartar
  2 tablespoons edible oil
- Add colouring if you wish.
- Warm ingredients in a pan gently until they thicken.
- Cool and store in a sealed container.

Plaster of Paris

- Prepare the mould or item to be copied by covering with Vaseline.
- Steadily sift Plaster of Paris into water until powder no longer sinks.
- Add a little more powder and stir to a thick, creamy consistency.
- Pour the plaster into the ready-prepared mould or over the item.
- After the Plaster of Paris has set hard, shape the edges with a knife and sandpaper smooth.
- Paint and/or varnish.

Note: For quicker setting add salt; for slower setting add vinegar.

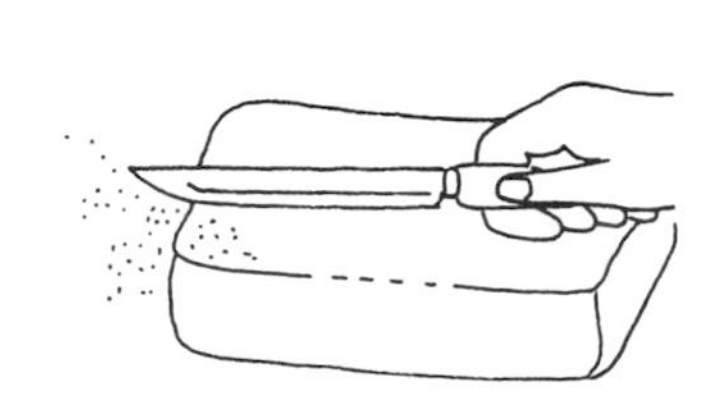

Paste and sand 'cement'

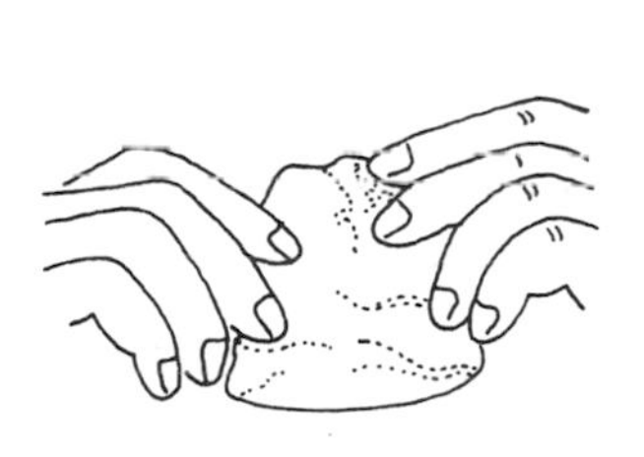

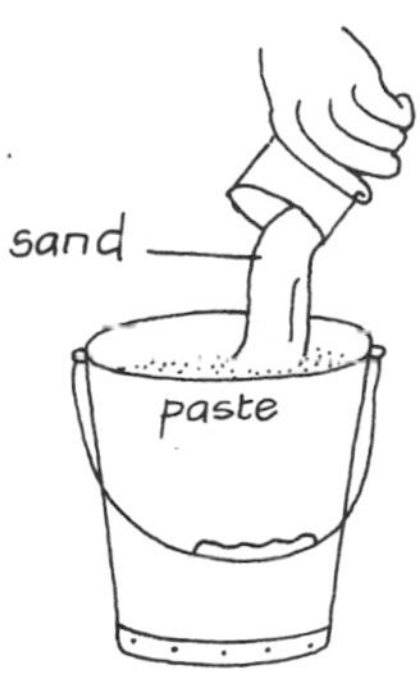

- Mix evenly together dry sand and flour paste (see page 118) or commercial glue.
- The wet cement moulds very easily and dries hard.

Paste and sawdust 'cement'

- Prepare as paste and sand, but use sawdust instead of sand.

Collecting and displaying

Pitfall traps

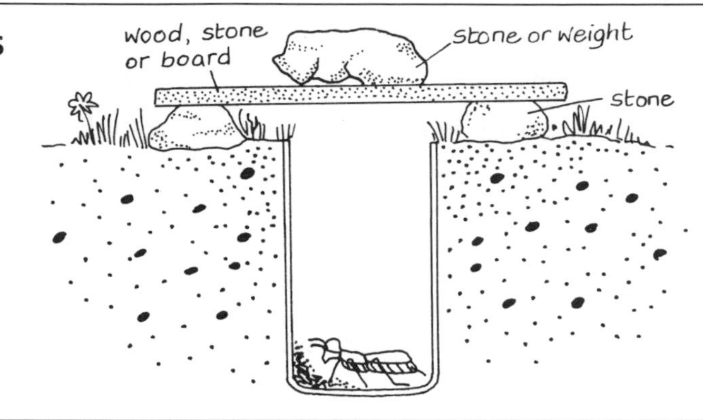

- Make a few holes in the bottom of a tin to let water escape.
- Bury the tin up to its rim in the soil.
- Cover the tin to keep out rain.
- Try out different types of food as bait.
- Check the trap regularly and remove it when finished with!

Soil life

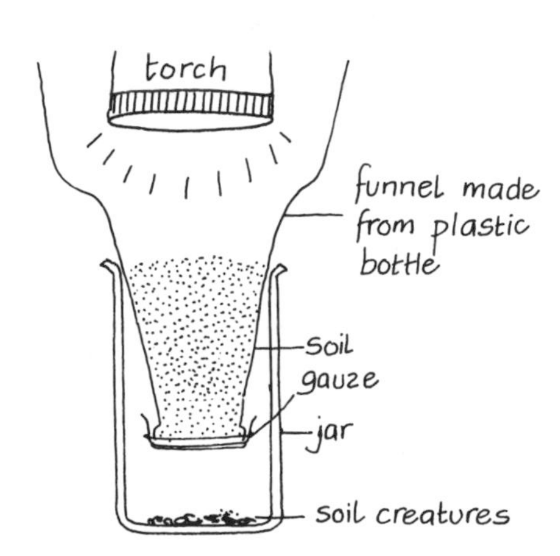

- Collect a sample of soil and place it in a funnel with a piece of gauze across its neck.
- Shine a bright light down onto the soil.
- Soil organisms usually prefer dark, damp and cool conditions so the heat and light drives them downwards until they drop into the collecting jar.
- Return organisms to the soil after examination, as many may dehydrate and die.

Collecting nets

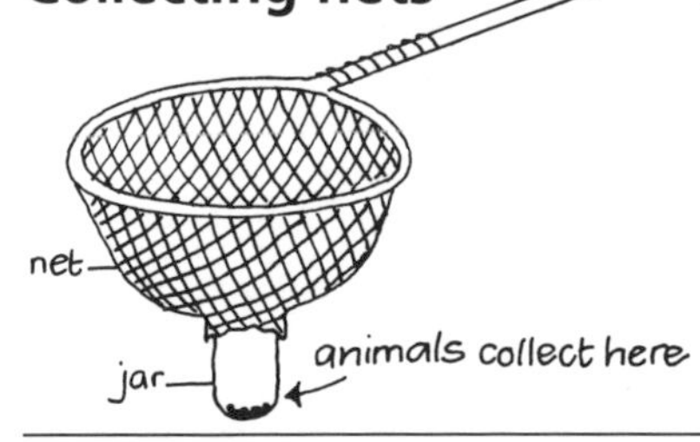

- Collecting nets can be made easily from sticks, some wire and mosquito netting.
- For collecting small water creatures use a fine net with a small jar attached to the blind end as shown.
- River nets can be used to catch small animals disturbed from stones and mud by a stick.

Collecting and labelling

- Wherever possible try to identify any plants or animals you collect.
- Leaves and flowers can be dried in a press and then fixed to paper with glue or tape.
- A thick loose-leaf file may be best for final display of plant specimens.
- Try to label all specimens as shown below.

  SPECIMEN: *leaf*
  NAME: *mango*
  LOCAL NAME: *mwembe*
  LOCALITY: *Mzumbe, Morogoro*
  HABITAT: *Mihigo's garden*
  DATE: *2nd August 1993*

Studying and displaying

Here are a few suggestions which may be useful if you wish to study animals over a period of time.

Worm jar

- Fill a plastic or glass vessel with soil and add the worms.
- Wrap black or dark paper around the jar to keep light away from the burrowing worms.
- Remove the paper to reveal the burrows.
- Make sure the soil is kept moist and never dries out.

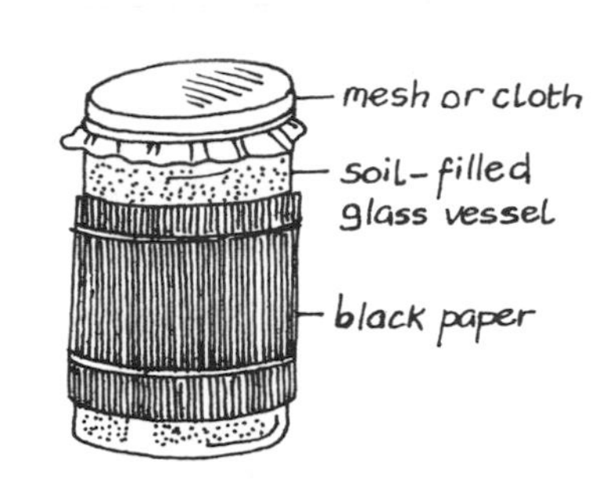

Flying insect cage

- Insects can be kept in many types of cage.
- Mosquitoes and other insects benefit from having water, vegetation and room to fly. The cage shown provides all these.

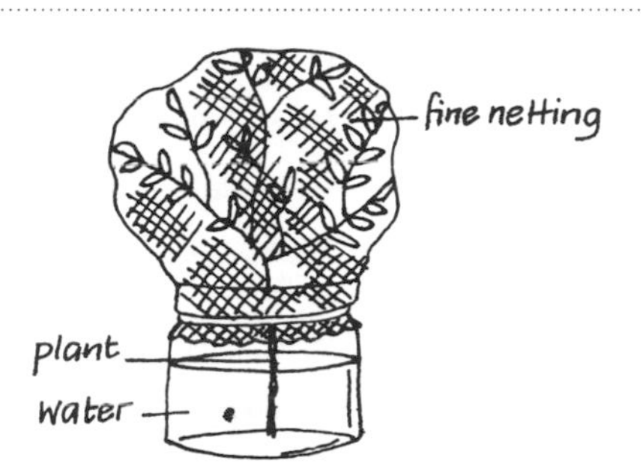

Reptile cage

- What would you need to add to this jar to make it suitable for keeping and observing lizards or other reptiles?

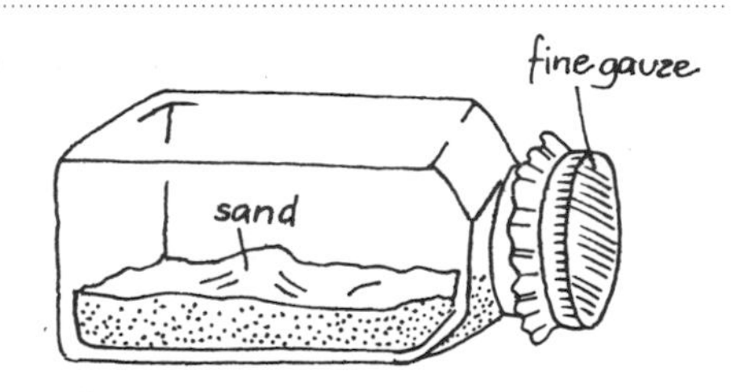

Aquarium box

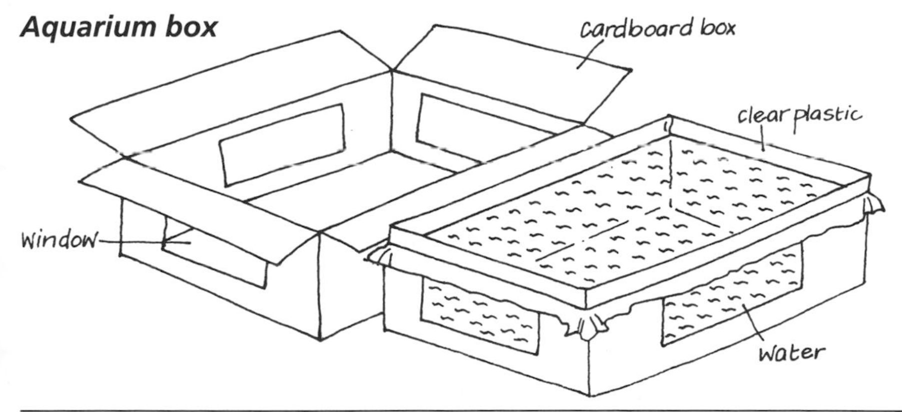

- Cut viewing windows in the sides of a box.
- Line the box with a large sheet of transparent plastic and fill it with water.
- Attach the plastic firmly, making sure it does not slip down from around the rim of the box.

Caring for animals

- Always treat animals with care.
- Some animals are dangerous, some scare easily.
- After study return animals to the place you found them.

Storage

Card and picture boxes

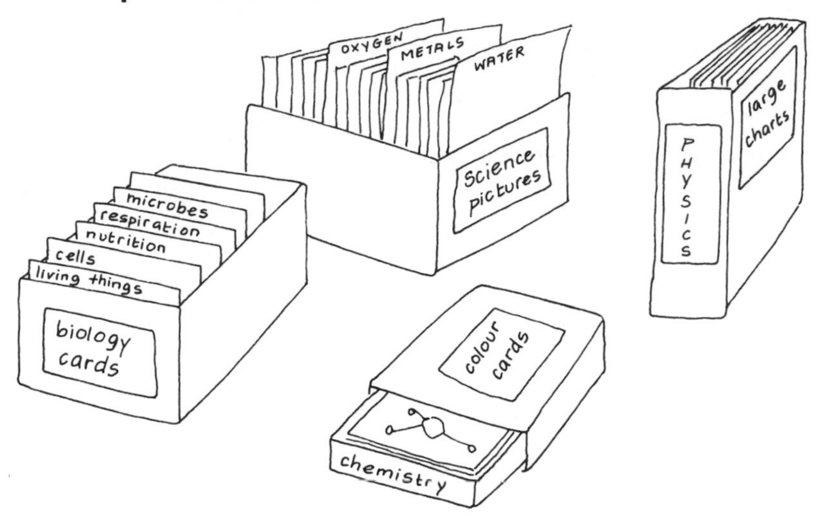

- Select suitably-shaped boxes.
- Cards and pictures can be stored in all sorts of boxes. Store according to syllabus topic or alphabetically.
- Dividers and compartments can be made from cardboard.

Matchbox drawers

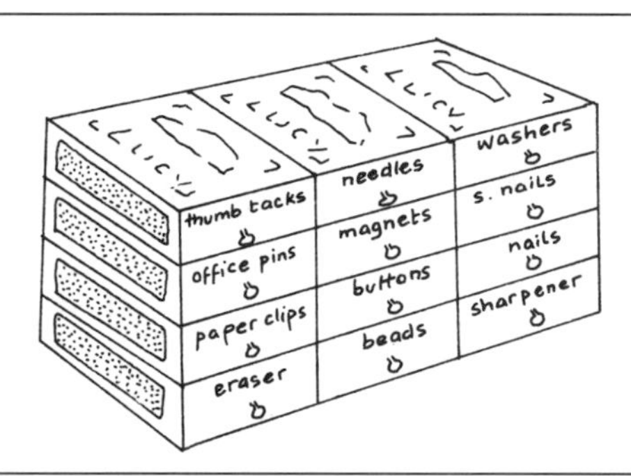

- Drawers to store small items can be made from matchboxes glued together as shown.
- Small pieces of string, wire or buttons can be used as handles.

Dividing boxes

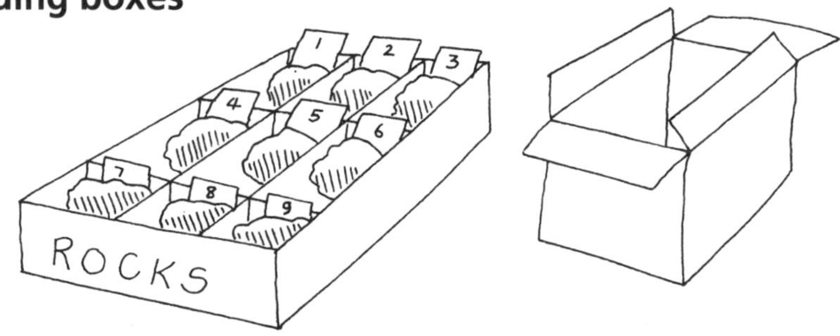

- Cut down the sides of boxes for displays.
- Samples can be sorted, then displayed or stored in cardboard boxes as shown.
- The flaps from the top of the box may be cut off and used as dividers for the same box.

Envelopes and bags

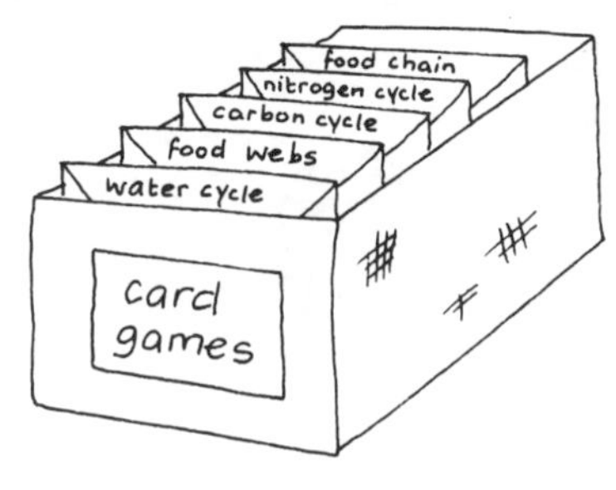

- Envelopes and bags of different sizes can be used for storage. Clearly label all containers.

Tins, cups and bottles

- Tins, cups and bottles make good storage containers. Some items do not need lids, others do.

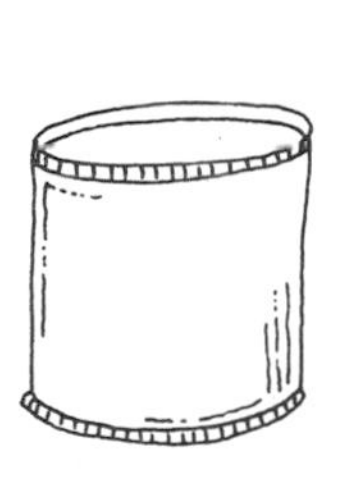

Reagent and Winchester bottles

- Winchester bottles can be replaced by any big bottle, but check that the container does not dissolve in the substance first! Dangerous substances, such as acids and alkalis, should always be kept in strong containers such as glass or suitable plastic.
- Reagents can safely be stored in used jam jars, perfume and scent bottles or film containers.
- Always identify the contents with a clear label.
- Search the market for ideas.

Folding and storing posters

- Make your posters fold into the size of a book. It is easier if the poster is made from paper of the right size.
- After use, fold the poster along the original 'book folds' and store in a cover.

Local sources of chemicals

Common name	Source	Chemical name
alumina bauxite	local quarry	aluminium oxide
aspirin	chemist	acetyl salicylic acid
baking *soda*	shop/market	sodium hydrogen carbonate (common name: sodium bicarbonate) Baking *powder* also contains sodium bicarbonate, but it may also contain other substances, e.g. citric acid, which may interfere in some experiments
bitter salt	chemist	magnesium sulphate
bleaching powder	shop/market	calcium chloro-hypochlorite
borax	chemist	sodium tetraborate – good reducing agent
burnt ochre, burnt ore, Indian red, Venetian red	all dyes – market	iron(III) oxide
calcite		mineral calcium carbonate
marble/limestone	eggshells, coral, dustless chalk	calcium carbonate
iron mordant	chemist	iron(III) sulphate
mothballs	shops, market	naphthalene
pearl ash	(rare and valuable)	potassium carbonate
oil of bitter almonds	flavouring – market/shop	benzaldehyde
salt	market/shop	sodium chloride
sylvine	(rare and valuable)	potassium chloride
spirit of wine	wine – shops	ethanol
silica/sand	beach	silicon dioxide
washing soda	shops/market	sodium carbonate
milk of magnesia	chemist	magnesium hydroxide
Chile saltpetre	fertiliser	sodium nitrate
cane sugar	crushed sugar cane	sucrose
grape sugar	crushed grapes (use the solution)	glucose
malachite	mine/quarry	copper carbonate
gypsum	chemist	calcium sulphate
vinegar	shop/market	ethanoic acid
lemon/orange/lime juices	shop/market	citric acid
malaria tablets (not all)	chemist	quinine
butane gas	shop	Trade names: Calor, Afrigas, Nidogas
some types of antiseptic	chemist	hydrogen peroxide, iodine
steel wool, steel screws, packing-case binding strips	shop/market/hardware	steel
staples	office/shop	tin plate
metal generally	garage/workshop	iron/steel etc.
electrical wires	shop	copper – remove insulation
kitchen foil/some drinks cans	shop/market	aluminium

The VSO ECOE Programme

Evaluating and Communicating our Overseas Experience

The need

Over the past 35 years more than 20 000 volunteers have worked overseas with VSO. Currently, there are over 1600 volunteers working in over 50 countries in Africa, Asia, the Pacific and the Caribbean for periods of two years or more. We have become increasingly aware that much of this valuable experience has not been recorded in ways which make it accessible and communicable. The ECOE Programme addresses this need.

The aim

The aim is to record volunteers' professional experience in books, reports, videos, seminars, conferences etc. This body of knowledge supplements and supports the work of individual volunteers. It also provides information which is accessible not only to volunteers but also to their employers overseas and to other agencies for whom the information is relevant. Care is taken to present each area of volunteer experience in the context of current thinking about development so that VSO both contributes to development discussions and learns lessons from them for the continuance of its work.

Advisory panel

A panel of opinion leaders in relevant professions and development thinking advises on the selection and commissioning of ECOE publications.

Publications

Agriculture and natural resources – A manual for development workers, Penelope Amerena, ISBN: 0 9509 0503 8, £9.95(excl. p+p)

Children actively learning – The new approach to primary eduation in Bhutan, Peter Collister and Michael Etherton, ISBN: 1 85339 111 5, £4.95 (excl. p+p)

Culture, cash and housing – Community and tradition in low-income building, Maurice Mitchell and Andy Bevan, ISBN: 1 85339 153 0, £6.95 (excl. p+p)

Introductory technology – A teacher's resource book, Adrian Owens, ISBN: 1 85339 064 X, £9.95 (excl. p+p)

Made in Africa – Learning from carpentry hand-tool projects, Janet Leek, Andrew Scott and Matthew Taylor, ISBN: 1 85339 214 6, £4.95 (excl. p+p)

Using technical skills in community development, Jonathan Dawson, edited by Mog Ball, ISBN: 1 85339 078 X, £4.95 (excl. p+p)

Water supplies for rural communities, Colin and Mog Ball, ISBN: 1 85339 112 3, £5.95 (excl. p+p)

To order these titles and for more information contact the ECOE Programme Manager,
VSO,
317 Putney Bridge Road,
London,
UK.
Telephone (+44) 081 780 2266.
Fax: (+44) 081 780 1326.

Index